The Future of Human Space Exploration

Giovanni Bignami • Andrea Sommariva

The Future of Human Space Exploration

palgrave
macmillan

Giovanni Bignami
IASF-INAF
Milan, Italy

Andrea Sommariva
Milan, Italy

ISBN 978-1-137-52657-1 ISBN 978-1-137-52658-8 (eBook)
DOI 10.1057/978-1-137-52658-8

Library of Congress Control Number: 2016936718

Printed on acid-free paper

This Palgrave Macmillan imprint is published by Springer Nature
The registered company is Macmillan Publishers Ltd. London

Foreword: The Future of Human Space Exploration by Johann Dietrich Woerner, General Director of the European Space Agency

In *The Future of Human Space Exploration*, Giovanni Bignami and Andrea Sommariva present a discussion of a very challenging topic: that of exploration. For the space community, exploration lies at the very core of space activity and brings together technology and science at the highest level. Exploration, whether robotic or human, requires that we think about possible effects that extend far beyond our day-to-day experience. The book focuses on all aspects of exploration, ranging from history, launcher and propulsion systems and spacecraft, to possible activities beyond science such as mining in the universe. It also covers the crucial aspects of human spaceflight. Large parts of the book are written in such a way as to give even the less-informed reader a clear understanding of the subject. At the same time, those already with an in-depth understanding can find detailed information about the physics that provides the background to these activities.

The most challenging part of exploration is neither the technology nor the science. Rather, it is to convince those organisations that handle public money to invest in exploration. The problem is that exploration is looking into the unknown and therefore cannot seriously promise a direct return on investment on Earth. However, examples from the past show that all exploration missions have had additional effects beyond their initial scientific purpose. For example, it was investigations of Venus that disclosed the greenhouse effect on Earth (better known as "climate change"). For the Rosetta mission, a special camera was developed that was able to differentiate between different shades of grey on the comet. This technology is now used for early forest-fire detection by observing forests and indicating smoke vapour or fog. In addition, exploration missions are usually organised in an international framework able to bridge earthly conflicts. A first-rate example of this is the International Space Station in which astronauts and cosmonauts from the USA, Russia, Canada, Japan and Europe all work together very successfully.

However, as such results are not the main objectives of exploration, there is a constant need to continue making the case for it.

Contents

List of Figures

List of Tables

1

Introduction

Earth is the cradle of humanity, but one cannot live in a cradle forever.

(Konstantin E. Tsiolkovsky)

Space exploration has always fascinated humankind. It has inspired the works of philosophers such as Lucretius, Kepler, and Kant and modern works of fiction by Jules Verne, Isaac Asimov, Arthur C. Clark, Fred Hoyle, and Italo Calvino. These works are not only pleasant entertainment but also ways to expand our imagination. They allow us to explore human responses to future scientific developments and to speculate on how they might develop. In the second half of the twentieth century, space exploration moved away from the realm of pure imagination. Sputnik and Soviet astronauts (beginning with Yuri Gagarin) orbiting the Earth and American astronauts landing on the Moon created an atmosphere of optimism. Optimism pervaded science, the public, and the arts, as Stanley Kubrick's film *2001: A Space Odyssey* explored the possible origins and fate of humankind in space.

After the optimism of the 1960s, human space exploration entered a state of flux. Humanity's presence in space centred on suborbital flights

G. Bignami, A. Sommariva, *The Future of Human Space Exploration*,
DOI 10.1057/978-1-137-52658-8_1

and short-term residence on the International Space Station. Paraphrasing Tsiolkovsky, humanity still lives in the cradle. Space agencies have been conducting robotic space exploration with great success. Many think of scientific research as space exploration's main goal. They are losing sight of other equally important goals: those of an economic, commercial, or cultural nature.And, in the longer term, spreading out into space may perhaps guarantee the survival of the human race. As physicist Stephen Hawking once said:

> The long-term survival of the human race is at risk as long as it is confined to a single planet. Sooner or later, disasters such as an asteroid collision or nuclear war could wipe us all out. But once we spread out into space and establish independent colonies, our future should be safe. As there isn't anywhere like Earth in the solar system, we would have to go to another star.

This book develops a scenario for human space exploration. Scenarios are not forecasts but rather ways to understand the dynamics shaping the future by identifying the primary driving forces at work today. Analyses of such scenarios do not rely on extrapolations from the past. They consider possible developments and turning points, which may or may not be connected to the past. Scenarios allow for qualitative changes not included in quantitative extrapolations of past trends. This is particularly important when analysing scientific and technical progress. Often, forecasts of these events are made obsolete by unpredictable innovations and scientific breakthroughs.

Our scenario examines: (i) the history of human space exploration from its beginning up to now; (ii) its short to medium-term prospects; and (iii) the possible longer-term developments. History helps us to understand better the motivations and constraints—technical, political, and economical—that shaped space exploration. The short to medium-term prospects enable us to identify the driving forces that will shape its next phase. While science and technology define the limits of what is possible: transforming these possibilities into reality depends on the economic and political benefits resulting from human space exploration. The economic benefits fall into three categories: direct effects measured by revenues generated by using space resources and related services and products; consumer welfare effects measured by the benefits to consumers beyond the value they paid for those products and services; and economic effects that arise from the efficiencies generated by those products and services.

Political benefits come in terms of international cooperation and world political stability. International cooperation is a necessary condition for human space exploration: the financial resources required for it are too large for any single nation to afford. If private companies are to seize the opportunities arising from space exploration, certainty over property rights and the uses of space resources are needed. This would involve extensive international cooperation. But we are struggling to find ways to cooperate. We are far from an advanced level of civilization in which international relations are solely based on cooperation and not conflict. Tensions arising from the control of natural resources, economic inequalities, and racial and religious conflicts are among the obstacles to international cooperation. Doubts exist whether we will be able soon to achieve such a civilization.

Despite this note of pessimism, the situation may not be hopeless. Michael Tomasello[1] argues that *Homo heidelbergensis* developed abilities to cooperate through mimicry and gesturing in the search for food. Later on, *Homo sapiens* expanded cooperative capabilities through common cultural backgrounds centred on conventions and norms. Through these arguments, Tomasello proposes that humans can continue to develop because they can absorb and share knowledge. The human race has thus the ability to find solutions to problems it has itself created, and, through this process, to reach the stars.

To ease reading, all technical details have been placed in footnotes. However, it is not necessary to read these in order to understand the text; they are there for the curious reader who wishes to know more.

Reference

Tomasello M. A natural history of human thinking. Cambridge, MA: Harvard University Press; 2014.

[1] M. Tomasello (2014).

2

Stepping Out of the Cradle: The Exploration of the Solar System from the 1950s to Today

Astronautics is the theory and practice of navigation beyond the Earth's atmosphere. Isaac Newton established the mathematical basis of astronautics in his treatise *The Mathematical Principles of Natural Philosophy*. They are embedded in his laws of motion and gravitation. The reactions in a spaceship's engine produce enormous pressures. They cause the expulsion of gas and/or radiation at high speed in the direction opposite to travel. It is this reaction force that pushes forward the engine and the spaceship attached to it.

Although Newton laid the mathematical foundations for it long ago, astronautics became a science in its own right in the early twentieth century. Starting in 1883, Konstantin Tsiolkovsky theorized many aspects of space flight. He published his most famous work, *The Exploration of Cosmic Space by Means of Reaction Engines*, in 1903. In this book[1] Tsiolkovsky derived the basic formula for rocket propulsion. This formula calculates the final velocity of a rocket from the escape speed of the gases and the initial (including propellant) mass and final (without propellant) mass of the spaceship. In other theoretical works, he studied

[1] Tsiolkovsky K (1995).

G. Bignami, A. Sommariva, *The Future of Human Space Exploration*,
DOI 10.1057/978-1-137-52658-8_2

gyroscopes and liquid fuel rockets; he calculated the escape velocity from a gravitational field; and he analysed the problem of the control of a rocket that moves between gravitational fields.

Tsiolkovsky elaborated the theory of space flight as a supplement to his philosophical inquiries on the cosmos. His works include speculative concepts, such as the industrialization of space and the exploitation of the natural resources to be found there. Indeed, he was the first theorist to support human space exploration. His works influenced generations of scientists and astronautical engineers from Europe, Russia, and the United States. During the twentieth century, advancements in astronautics and astronautical engineering[2]—the practice of navigation beyond the Earth's atmosphere—made possible the exploration of the solar system, including human space exploration. As Wernher von Braun once said: "The rocket will free man from his chains: the chains of gravity that still tie him to this planet. The gates of heaven will then be open."

Below we will retrace the steps of human exploration of the solar system up to the present day. This will help us to understand the motivations and the constraints that shaped it. Analyses of political, economic, and cultural environments also help us to glimpse the short to medium-term developments.

2.1 The Golden Age of Human Space Exploration: 1957–1973

Human space exploration reached its peak in the late 1960s and early 1970s, driven by competition for power among nation-states. Under pressure of geopolitical competition, the United States and the Soviet Union created vast national space programmes. Sputnik and an orbiting capsule

[2] Pioneers conducted experiments during the first three decades of the twentieth century, such as R.H. Goddard in the United States; Hermann Oberth, Willy Ley, and Wernher von Braun in Germany; and Sergey Korolyov in the Soviet Union. Astronautical engineering progressed in the decades following the end of World War II. It is also worth mentioning the progress made in the field of the dynamics of orbital motion applied to calculate spaceships' trajectories and orbital manoeuvres (see Shoemaker and Helin 1978) and the use of gravity to accelerate them (Michael Minovitch 1961).

with an astronaut (Yuri Gagarin) created panic in the United States. In response to the challenges posed by the Soviet Union space programmes, the American military establishment proposed that it take the lead in meeting these challenges. But President Eisenhower, fearing the entry of the military-industrial complex, decided that "space was to be used only for peaceful purposes." He proposed the National Aeronautics and Space Administration (NASA) to Congress, which it promptly approved. NASA thus became the titular owner of all US space programmes.

In 1961, the young President Kennedy gave his famous speech to Congress, which can be summed up in three words: "man, moon, decade." But neither the President nor NASA knew how to achieve that goal. The staff at NASA were in a complete panic, but the pride inherent in fulfilling their President's commitment was a potent stimulus. Although Americans did not embrace von Braun because of his Nazi Party membership, he did become the central figure in the Apollo programme. Under his guidance, assisted by engineer Rocco Petrone,[3] NASA developed the Saturn V rocket, a modified version of the Jupiter rocket. Von Braun later called the Jupiter "an infant Saturn." Saturn V became the cornerstone of the Apollo programme, and it remains today the most powerful rocket ever built, and the only one to have carried humans beyond Earth orbit.

At first, von Braun thought of assembling various modules (the command/service module and the lunar excursion module) in Earth orbit. Space shuttles would bring the modules into orbit. A second Saturn rocket would then take them to the Moon; however, two Saturn V rockets would increase costs and might cause delays in the mission. So NASA engineer John Hubolt suggested the lunar orbit rendezvous (LOR) mission mode.[4] With LOR, a Saturn V rocket first inserts the modules into

[3] Petrone graduated from the US Military Academy at West Point in 1946. He earned a master's degree in mechanical engineering from the Massachusetts Institute of Technology in 1951. His career in rocket development began in the early 1950s in Huntsville, where he assisted in the development of the Redstone rocket. He served as director of launch operations at NASA's Kennedy Space Center from July 1966 until September 1969 and afterwards as Apollo programme director. His NASA colleagues described him as demanding. Clearly he played, along with von Braun, a vital role in the success of the Apollo missions.

[4] LOR was not new. The first mention of it dates back to 1916 when Yuri Kondratyuk, a Ukrainian engineer, proposed it (see Wilford J. 1969). His analysis identified LOR as the most economical way of landing a man on the Moon.

low Earth orbit and then propels them into lunar orbit using the last stage of the rocket. The lunar landing mission starts from lunar orbit, so only one rocket is used instead of two. A fierce debate convinced von Braun and NASA to adopt this solution in 1962. Undoubtedly, the decision to use LOR was vital to the manned landing on the Moon by the end of the decade.

This solution saved time and money, though NASA paid a high cost: it discarded the assembly of the modules in low Earth orbit and the related infrastructure, which constituted the core of a space station. This had important consequences for longer term plans for human space exploration. To meet the schedule imposed by Kennedy, NASA lowered the safety standards of the missions. This eventually caused headaches. During the first lunar landing mission, the on-board computer failed. With Neil Armstrong manually piloting it, the landing module was set down on the Moon with only a few seconds of fuel remaining. Even more serious was the case of the Apollo 13 mission, which took place a year later: an oxygen-tank explosion made the service module unusable during the cruise phase to the Moon. The astronauts instead occupied the lunar landing vehicle, uncomfortable but serviceable, and so returned safely to Earth. In retrospect, it was only with considerable good luck that, between July 1969 and December 1972, 12 American astronauts set foot on the Moon.[5]

Kennedy's 1961 speech led the Soviet Union to create its own programme[6] for landing astronauts on the Moon. But the programme suffered from two disadvantages. The first was the slow development of the computer industry in the Soviet Union; the second resulted from the division of work among several institutions: design bureaus which competed with each other, unlike the Apollo programme which had one coordination centre, NASA. But Sergiei Korolyov's design bureau enjoyed a privileged position due its successes with the orbiting Sputnik and the first manned space capsule. He proceeded with two missions: a cislunar one and a manned landing on the Moon. The first programme saw the space capsule Soyuz 7 K-L1 launched by the UR-500 K rocket

[5] In depth analysis of the Apollo programme is provided in Launius (1994).

[6] The Soviet space programme history is told by I. Shklovsky (1991).

(Proton). A second one envisioned the same space capsule launched by an N-1 rocket. The Soviet Union cancelled the first programme in 1970 and the second one in 1974. The success of the Apollo mission and a series of launch failures of the N-1 rocket were the main reasons behind these cancellations.

Both programmes enjoyed a high level of support among their respective populations and received large funding. Opinion polls conducted in the United States showed a consensus of over 60 % of the population. Because of the closed nature of Soviet society, we have only anecdotal evidence on the consensus among the Soviet population, though it reveals deep and genuine support. The Apollo programme cost $22 billion (equal to $153 billion in 2013). It accounted for around 4.5 % of the total expenditure of the federal government in 1967. While there are no official statistics of costs incurred by the Soviet programme, American and British intelligence services offered a few estimates. They suggested figures close to those of the Apollo programme.

The economic benefits of these programmes were large. The multiplier effect stimulated the United States economy: for every dollar spent on the Apollo programme, at least $3 were fed into the economy.[7] Many inventions emerged as by-products of the Apollo missions, and the economic impacts of technological innovation predominated over the medium to long term. The financial support of the US government to research and development contributed to the rapid growth of the semiconductor and the computer industries. The demand for these products by space and military programmes helped the development of these industries. However, despite the obvious economic benefits, human space exploration came to a halt after the American astronauts' landings on the Moon.

2.2 The Missed Opportunity

Von Braun developed a fascination for interplanetary flight when he was at school in Germany. During the war and in the late 1940s, he used his spare time to write a science fiction novel on a manned mission to

[7] See Chase Econometric Associates (1975) and H. R. Hertzfeld (1998).

Mars. He based his story on comprehensive engineering diagrams and calculations, which were included in an appendix.[8] The mission included seven passengers and three cargo ships, and made a one-year round trip to Mars. The German space flight journal *Weltraumfahrt* published the appendix in a special edition in 1952. Later that year, Umschau Verlag in West Germany published the book as *Das Marsprojekt*. Henry J. White translated it into English. The University of Illinois Press published it under the title *The Mars Project* in 1953.

During the 1960s, von Braun worked out a plan, within NASA, for the human exploration of Mars. He tweaked his original Mars expedition scenario to halve the manned spaceship's size to under half the mass of Boeing's 1968 Integrated Manned Interplanetary Spacecraft (IMIS). He contemplated the launch of 16 automated cargo spaceships and two nuclear-powered manned spaceships. This would enable the two spaceships to travel in convoy, providing back-up and mutual support. These spaceships were planned to be reusable for future expeditions. The following elements characterise the manned spaceships:

- Two lateral modules fuelled for changes in velocity to direct the spaceships toward Mars. They would separate and manoeuvre back to rendezvous and dock with the space station. There, they would refuel. They could then be reused for Earth–Moon shuttle services or future Mars expeditions.
- One central module powered by a nuclear engine developed in the Nuclear Engine for Rocket Vehicle Application (NERVA) project. This would provide the thrust for: insertion into Mars orbit, the return trip to Earth, and necessary manoeuvres to re-enter Earth orbit. Significant fuel savings would result due to insertion in a highly elliptical orbit around Mars.
- A habitable module for six astronauts in each spaceship. Each would have resources for 12 astronauts, if one spaceship needed to be abandoned during the mission.

[8] The appendix to *The Mars Project* contained technical specifications for the mission to Mars. Von Braun envisioned a scientific expedition involving a fleet of ten spaceships with 70 crew members. The spaceships were to be assembled in Earth orbit using materials supplied by reusable space shuttles. He planned to use Hohmann trajectories to transport the spaceships from Earth orbit to Mars orbit. He calculated each ship's size and weight, the fuel they needed for the round trip, and each rocket's burn to effect the required manoeuvres.

- One excursion module in each spaceship. The modules would descend to Mars from a highly elliptical orbit. They would be able to accommodate six astronauts for a 60-day stay on Mars. Since the spaceships flew in convoy, two modules could descend onto Mars next to each other. Six astronauts would subsequently be able to use one module to rejoin the orbiting spaceships at the end of their Mars mission.
- Sixteen robotic vehicles sent in advance to Mars. They contained the equipment and propellant needed for the return to Earth. On the return trip, 12 vehicles would transport samples collected during the mission. The remaining four could be redirected to enter Venus's atmosphere.

The project envisioned a fleet of shuttle vehicles ferrying parts to be assembled in Earth orbit. Von Braun contemplated two types of missions: one with a short duration and the other with a long duration. These missions are called the conjunction and opposition classes.[9] Conjunction-class trajectories are typified by low-energy transfers and by a long stay on Mars, around 500 days. The long stay provides the time required by Mars and Earth to reach relative positions to make possible the minimum energy transfer for a return. Opposition-class trajectories are characterised by one low-energy and one high-energy return transfer, and by around 40 days stay on Mars. Since an opposition-class expedition requires more energy, it demands more propellant. In fact, it needs ten times the propellant mass than does a conjunction-class trajectory. The proposed mission profile with a short stay on Mars was:

- 12 November 1981: insertion in a trajectory to Mars. Each spaceship's mass at the start was to be 727 tons. At the detachment of the two lateral modules, the mass of each spaceship would be reduced to 614 tons.

[9] The terms conjunction class and opposition class refer to Mars's position relative to Earth. In the former, Mars moves behind the Sun as seen from Earth halfway through the expedition. In the latter, Mars is opposite the Sun in Earth's skies at the expedition's halfway point.

- 9 August 1982: insertion into Mars orbit. The mass of each spaceship before insertion was estimated at 295 tons and, after a successful insertion, 226 tons.
- 10 August 1982: the excursion modules disengage from the spaceship and head to the surface. Meanwhile, the astronauts remaining on board the orbiting spaceship analyse the surface of Mars and its two moons.
- 28 October 1982: the manned spaceships are inserted into their return trajectory to Earth. The mass of each spaceship at the start is 172 tons.
- 28 February 1983: Venus flyby. This reduces the velocity of the spaceships. It enables the launch of the remaining robotic spaceships into the atmosphere of Venus to perform scientific experiments.
- 14 August 1983: entry into Earth orbit and docking with the space station. The final mass of each spaceship is 72.6 tons.

With hindsight, von Braun's programme was too ambitious because of the state of technology of nuclear propulsion at that time. But it remains a cornerstone project for the human exploration of the solar system. Technically, it established two necessary conditions for interplanetary missions: first, the use of nuclear-powered spaceships to reduce travel time and diminish astronauts' and sensitive spacecraft systems' exposure to cosmic radiation; second, a space-port to assemble spaceships and to insert them in trajectories to target destinations. Although it is possible to conceive launching nuclear-powered spaceships from the Earth's surface, the former solution reduces security risks associated with the second one.

On 15 September 1969, the NASA Task Group approved von Braun's plan. In the following days, he released the final report to Congress, which did not approve it, although only by a few votes. One of the project's weaknesses was the mission's high costs. Von Braun operated under the impression that whatever financial resources NASA might need, the government would provide them. But the United States government had other urgent priorities. Military expenditure had precedence because of the expanding Vietnam War, plus the Cold War was reaching its peak. And yet winning the competition with the Soviet Union was the main goal of the Apollo programme. But once achieved, policy makers and opinion leaders wanted answers to tough questions, such as: What are the long term goals of human space exploration and the benefits resulting from these missions? Since there were no compelling comprehensive answers, political support vanished. With von Braun's latest dream cancelled, NASA sidelined him at its headquarters. He left NASA in 1972 and died in 1977. Following Congress's decision on the Mars mission, President Nixon recommended reducing NASA's budget. But if the truth be told, he was not a fan of space activities, and most likely he wished to contrast himself from his nemesis Kennedy. From 1970 onwards, NASA expenditure as a percentage of the total federal budget declined continuously, reaching 0.5 % in 2013, although, due to the shuttle programme, there was an exception between 1987 and 1993. The declining trend clearly indicated that no consensus existed among policy makers on human space exploration.

2.3 Waiting for the Great Leap Out of the Cradle

Between 1972 and today, robotic space exploration has prevailed over human space exploration. It has been a period of consolidation and incipient international cooperation by national space agencies due to the limited financial resources made available by governments. But human space exploration has remained a long term goal of a few space agencies, which have been allocating limited resources to developing new propulsion systems for human interplanetary missions. During this period, a space economy has also emerged. A space economy is the full range of activities and resources that create value while exploring, managing, and using space. The Organisation for Economic Co-operation and Development (OECD) defines the global space economy to include the space industry's core activities involving the manufacturing of space vehicles and satellite operations, as well as other space related activities. But these activities are carried out in Earth's vicinity. Recently, the allure of deep space has captured the minds of a new generation, though absent from this new trend is the kind of strong government directive seen in the Apollo era. In its place, we observe the awakening of individuals and private industries, who realise the possibilities of the democratisation of space and the expansion of humankind's economic sphere.

2.3.1 Space Exploration Extends to Many Nations

A growing number of countries are involved in space activities. These endeavours include: building rockets and spacecraft, designing experiments to go on board spacecraft, and providing and training the astronauts. Few countries can put vehicles into orbit around Earth. To date only 14 countries have production and orbital launch abilities. These are Russia, the United States, nine European countries through the joint effort of the European Space Agency (ESA) and Arianespace, China, Japan, and India.

During the 1950s and 1960s, Europe and Japan were rebuilding their economies after the catastrophes caused by World War II. Consequently, they were absent from space exploration activities. In the late 1960s, several European countries looked enviously at the American space exploration programme and its effects on technological innovation. In 1975, a group of European countries created ESA, a body independent of the European Union. The main goal of ESA is to enable Europe to become and remain competitive in space technology, and to carry out scientific research. ESA coordinates space programmes for the participating European countries. ESA's space flight programme includes human space flight, mainly through participation in the International Space Station. The main European launch vehicle, Ariane 5, is operated through Arianespace with ESA sharing in the costs of launching and developing launch vehicles. It maintains a major space port, the Guiana Space Centre at Kourou, in French Guiana.

ESA is the first example of international cooperation on a regional level. Its activities are grouped into two categories: mandatory and optional programmes. The programmes carried out under the general budget and scientific programmes are mandatory. These are funded by member states in proportion to the size of their gross domestic products.

Member countries are free to decide on their participation in optional programmes, which include projects such as Earth observation, telecommunications, space transportation, and manned space flight. For example, the International Space Station and research in microgravity are funded under optional programmes. ESA also relies on cooperation with NASA. But the changed circumstances in the 1990s (i.e. tough legal restrictions on information sharing by the United States military) led to decisions to depend more on itself and on cooperation with Russia.

China's space programme began in 1956 as part of Sino-Soviet cooperation. China began a rudimentary ballistic missile programme in response to perceived American threats. After the 1960 Sino-Soviet crisis, China continued its space programme independently of the Soviet Union. The main task centred on ballistic missiles. The Central Military Commission controlled this programme to meet national defence needs. As the space race between the two superpowers reached its climax with the conquest of the Moon, the Chinese government decided that they should not fall further behind. So China started its own crewed space programme. In the early 2000s, the Chinese Space Agency started an ambitious programme of human space exploration, having the Moon as a target. Accelerated programmes of technological development culminated in Yang Liwei's successful orbital flight aboard Shenzhou 5 in 2003.

In 1990, the Chinese government reorganized its space programme along with other defence-related industries, initially establishing two agencies: the China National Space Administration (CNSA) and the China Aerospace Corporation (CASC). The former was responsible for policy, the latter for execution. This arrangement proved somewhat unsatisfactory as these two agencies were in reality one large agency sharing both personnel and management. After reorganisation, the space programme became the responsibility of the CNSA and the Science, Technology, and Industry Commission. As part of a massive restructuring in 1998, the government split CASC into smaller state-owned companies. Government entities set operational policy. They contract out their operational requirements to entities that are government-owned, but not government-managed. Companies that produce launchers and satellites are under the control of the government, which permits them freedom in operations.

In late 1962, Japan also entered the realm of space exploration. The Japanese government created three agencies: the National Space Development Agency, the Japanese Aerospace Laboratory, and the Institute of Science Astronautics. Promoting space research and technological innovation in Japan were the main goals of these institutions. In 2003, the government merged the three agencies into a single Japanese Space Agency (JAXA). The entry of China into space exploration in the early 2000s was an important motivation for the creation of JAXA. Since then, JAXA has established cooperation programmes with ESA, NASA, and the Indian Space Research Organization (ISRO). JAXA handles research, technology development, and satellite launchings, and is involved in many advanced projects. These range from research and development of a space-based solar-powered satellite, asteroid exploration, and the possible manned exploration of the Moon.

The government of India founded ISRO in 1969 within the Department of Atomic Energy. In June 1972, the newly established Space Commission and the Department of Space took over the control of ISRO. Promoting space technologies and their applications for Indian economic development was the goal of this organisation. ISRO developed two rockets: the Polar Satellite Launch Vehicle for launching satellites into polar orbits, and the Geosynchronous Satellite Launch Vehicle for placing satellites into geostationary orbits. These rockets have already launched several communications and earth-observation satellites. In 2008, ISRO sent a mission to the Moon, Chandrayaan-1. In 2013, it launched the Mars Orbiter Mission (MOM), which is a "technology demonstrator" project to develop technologies for design, planning, managing, and operating interplanetary missions. It carries five instruments to advance knowledge of Mars, which is a secondary scientific goal. When MOM successfully entered Mars orbit on 24 September 2014, India became the first Asian country to do so.

Over the years, other nations have initiated space activities, although with smaller programmes. In 2013, the total budget of the main spacefaring nations[10] amounted to $76 billion (Fig. 2.1). This repre-

[10] European national budgets are separate from their contributions to ESA. Member states take part to varying degrees in the mandatory and optional ESA space programmes.

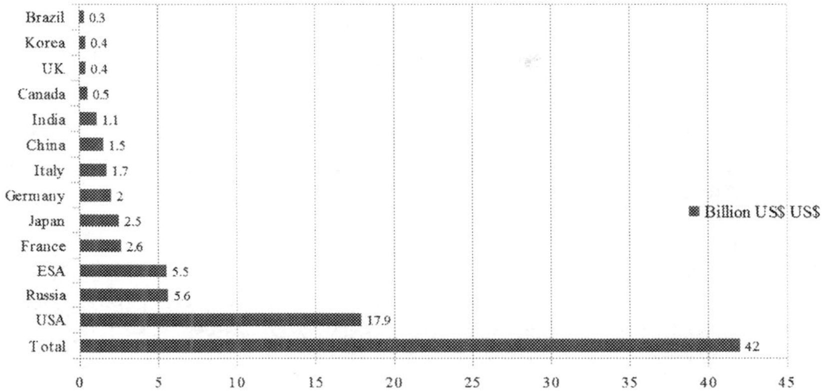

Fig. 2.1 Space budgets of the main spacefaring nations (*Source*: National space agencies' statistics)

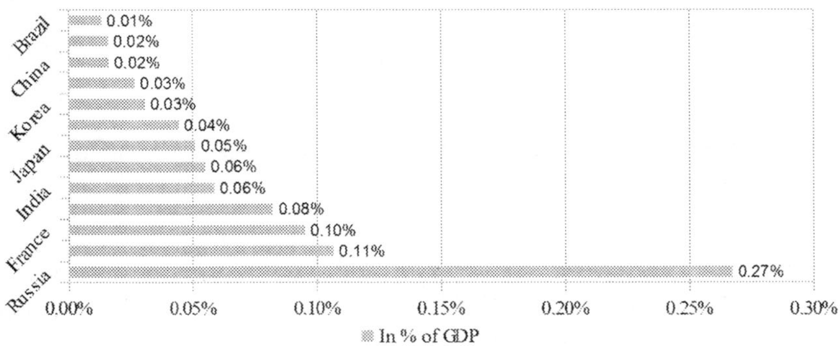

Fig. 2.2 Space agencies' budgets as a percentage of GDP (*Source*: National space agencies' statistics and IMF)

sented around 0.15 % of the listed countries' GDPs. The space agencies' budget of the United States, Russia, ESA, Japan, and China accounted for 89 % of the total.

While total dollars count, how much of a country's economy is invested in space activities is also relevant. Space agencies' expenditures as per cent of GDP shows the relative priority of space in overall governmental policies (Fig. 2.2).

By share of GDP, Russia takes the lead. This is followed by the United States, France, and Italy. India moves up the rankings ahead of Germany, Japan, ESA, Korea, and China.[11] India takes space activities seriously as a part of its economy. Finally, attention should be given to the shares of human space flight (HSF) within the total budgets of the national space agencies. NASA and the RKA (the Russian space agency) are unique in spending high amounts of their budgets on HSF, in a range between 35 and 45 %. Meaningful figures for historical or projected spending on HSF by China are virtually impossible to get, though its space-station programmes and manned-space flights show the high priority given by the Chinese to HSF. The Japanese, Canadian, and European shares for HSF in their total spending lie between 10 and 20 % and have even been declining. While India has ambitious plans, its HSF spending is still a tiny fraction (1 %) of ISRO's overall budget. The investment portfolio priorities of space agencies, except for NASA, the RKA, and the Chinese Space Agency, are centred on science projects, technology, and satellite activities.

2.3.2 International Cooperation: The International Space Station

The International Space Station (ISS) is a project jointly owned by NASA, the Russian RKA, the European ESA, the Japanese JAXA, and the Canadian CSA. Ownership and use of the space station is established by intergovernmental treaties and agreements. The history of the space station is a complex one. After the end of the Apollo programme, plans for building a space station in low Earth orbit vanished along with von Braun. Eventually, proposals for a large space station re-emerged in the United States. A space station had formed part of an effective strategy to bury the "evil Soviet empire" under the weight of US technology. This proposal saw a space station in an orbit inclined at 28° to the equator, that is the latitude of Cape Canaveral,

[11] Meaningful figures for historical spending on the Chinese space programme are virtually impossible to obtain. The estimates of civilian-space spending are derived from Western expert reports, showing that CNSA spending is around one-tenth of NASA spending.

the Shuttle home base. "Hawks" in the Reagan administration gave it a name: "*Freedom*." NASA realised the insufficiency of its own resources, and "*Freedom*" remained an expensive paper project. Meanwhile the Soviets built their own no-frills space station, consistent with their pragmatic space programme. The Mir station (*mir* in Russian means both "peace" and "world") was the first modular space station and was assembled in orbit from 1986 to 1996. The station served as a microgravity research laboratory. Crews conducted experiments in plant biology, human biology, and physics, and made observations in astronomy and meteorology.

After the Soviet Union's demise, Russia adopted a relatively cooperative foreign policy with the West. Despite economic turmoil, the country retained advanced technical knowledge of, and capabilities in, space operations. Since NASA was happy to work with Russia's space agency, politicians on both sides jumped on that wonderful opportunity. In 1993, Al Gore and V. Chernomyrdin signed an agreement to design, develop, and use a permanently inhabited civil space station for peaceful purposes. A whiff of pork hung over the plans as the ISS kept open factories that otherwise might have closed. Downstream of the political outcome, technical problems emerged. The inevitable price to pay for working with Russia was to build an ISS in an orbit inclined at 51° to the equator. To transfer from one orbit to another, a spaceship has to spend energy, that is has to burn more fuel. Celestial mechanics stipulates that it is easier to transfer from an orbit inclined at 28° to one inclined at 51° than the reverse. A spaceship can move from orbit "Cape Canaveral" to orbit "Baikonur," but the reverse process is inefficient. This meant that the shuttle would require more energy, though NASA thought they could afford it.

Plans, as always optimistic, saw the completion of the ISS by 2003. But it took an extra eight years to complete. Russia launched module "Zaria" ("dawn" in Russian) from Baikonur in 1998: indeed the first launches were all Russian. They used the cargo vehicle "Progress" to make 36 flights to the ISS. Later on, launches were split equally between the US Space Shuttle and the Russian spacecraft Soyuz. Recently, Europe and Japan have used their own cargo vehicles. Although, in the ISS global economy, cargo business is not a very important element, it has helped to justify ESA to operate its launch base in Kourou (French Guiana). For the first time, a European vehicle reached the ISS with authorisation to dock, having started from a space port other than Baikonur or Cape Canaveral. The Japanese have their own transport vehicle in place.

Since the ISS's overall costs amount to around $150 billion,[12] tough questions have been raised about its usefulness. The ISS has achieved nothing of the strategic significance imagined at the beginning, because the "Soviet evil empire" no longer exists. It is an inefficient base for launching spaceships to explore the solar system because of its high inclination to the ecliptic plane.[13] But does it have a scientific usefulness? Yes, but it is modest, and not enough to justify the large investments in scientific experiments as politicians have tried to make the world believe since the Reagan presidency. This has resulted in the justifiable critical reaction of the global scientific community: "not with my money," scientists said then and are saying today. The ISS costs cannot be justified by science alone, or, even less, to be attributed to scientific motives, since it was and is mainly a political programme. But certainly the investment in scientific experiments should continue to be made since the ISS is a fait accompli, and they are indeed being carried out.[14]

The ISS has also proven useful as a training ground for further human space flight. So far, several dozen astronauts from many countries have

---

[12] This includes: NASA's budget of $58.7 billion; Russia's $12 billion; Europe's $5 billion; Japan's $5 billion; Canada's $2 billion; and the cost of 36 shuttle flights to build the station, estimated at $1.4billion each, or $50 billion in total. Assuming 20,000 person-days of use from 2000 to 2015 by two to six-person crews, each person-day costing $7.5 million, the total is around half a billion dollars.

[13] If the Sun's path is observed from the Earth's reference frame, it moves around Earth in a path tilted at 23.5°. This path is called the ecliptic. Observations show that other planets, except for Pluto, orbit the Sun in the same plane. The ecliptic plane thus comprises most of the objects orbiting the Sun. A spaceship leaving the ISS for interplanetary missions will have to move from an orbit inclined at 51° to one inclined at 23.5° and so require higher fuel consumption. Hence, it is clear that the ISS is inefficient as a staging point for interplanetary missions.

[14] Researchers are studying the effects of weightless environment on evolution, development, growth, and internal processes of plants and animals. They propose to investigate the effects of micro-gravity on the synthesis and growth of human tissues and proteins. Other researchers are investigating the states of matter (in particular superconductors) outside the station at low temperatures.

visited it. This enables on-going studies on the effects on the human body of extended periods in space. Among others, these studies focus on muscle atrophy and bone loss due to low gravity. But these experiments do take place in low Earth orbit, that is above the atmosphere but within the Earth's magnetic field, which blocks the cosmic radiation that otherwise would be hitting the vehicles and the astronauts within. But the most serious problems for human interplanetary flight come from exposure to ionizing cosmic radiations from the Sun and the ones coming from deep space. In other areas, the ISS is a useful laboratory to test 3D printing[15] in weightlessness, which could be important for future space operations. It is also relevant for building in-space infrastructure, equipment, and other products using raw materials from space. Over the long run, the ability to deliver components on demand without the need of launch vehicles could redefine how space-mission strategies work.

There is no question that the ISS has taught the United States, Russia, Europe, and Japan to develop new technologies and has stimulated industrial and global collaboration. This is the ISS's most important role. Its construction, assembly, and operation require the support of facilities on Earth managed by the partner agencies and industries involved in the programme. The ISS programme brings together international flight crews; multiple launch vehicles; globally distributed launches, operations, training, engineering, and development facilities; communications networks; and the international scientific community. Whether this collaboration will continue is another story. Recent geopolitical tensions between Russia and the Western world have put international cooperation in space in jeopardy. The Russian space agency says it will support US plans to keep the ISS operating until 2024. Afterwards they plan to split off three still-to-be launched modules to form a new, independent orbital outpost. The decision by China to build its own space station may send us back to a world where geopolitical competition dominates national space exploration programmes. It is too early to say whether this scenario will materialise.

[15] Recently, SpaceX's CSX-4 vehicle brought to the space station a 3D printer, built by the company Made in Space. This started the first 3D printing in space. It took over an hour to finish the job. More experiments will follow in the following months.

Assuming continuous international cooperation, the final tough question is what to do after the ISS is "mothballed." As noted earlier, the ISS is not an efficient base to launch spaceships on interplanetary missions, while the same goes for China's space station whose orbit is inclined at around 47° to the ecliptic plane. Low earth orbit with smaller inclination or Lagrangian points between Earth and the Moon must be chosen. The latter solution is preferable. First, the space station would be where the Earth and Moon's gravitational influences cancel each other out, so that spaceships need no energy to break free from Earth's gravity. Second, for safety reasons, nuclear-powered spaceships can only turn on their engines at large distances from Earth.

The costs of building from scratch such infrastructure are so large that they will not be workable for any single nation. We are at a point in history where international cooperation or its lack can influence the dynamics of the development of future human space exploration. A proposed alternative[16] is to attach a VASIMR[17] drive module at the vacated node with its own on-board power source. This will let the space station or part of it be moved to a Lagrangian point between the Earth and the Moon. A low thrust applied continuously for days or months can move the ISS or part of it into lunar orbit without ripping apart its fragile structures. Once in the Lagrangian point, it will serve as the core staging post for future interplanetary missions. This alternative would enable us to recoup much of the investment made as it costs significantly less than building a new space station from scratch. But even in this case, international cooperation among spacefaring nations will be essential.

2.3.3 Development of New Propulsion Engines

Despite meagre resources, space agencies are financing research and development programmes on new propulsions schemes. The goals are

[16] Gene McCall, retired chief scientist for the US Air Force Space Command, advanced this solution in 2012.

[17] The Variable Specific Impulse Magneto-plasma Rocket (VASIMR) is an electromagnetic thruster. It uses radio waves to ionise and heat a propellant and magnetic fields to accelerate the plasma and so generate thrust. It is one of several types of electric propulsion.

shortening astronauts' time in space and reducing propellant mass. When discussing the efficiency of propulsion, engineers focus on how effectively it uses reaction mass. Reaction mass is the mass against which spaceships operate to produce acceleration. In a rocket, for example, the reaction mass is the fuel shot backwards to create the propulsion. Specific impulse, that is thrust deliverable by ejecting a unit weight of propellant, is one way of measuring propulsion efficiency. Why are we interested in specific impulse? First, it gives us a quick way to calculate the rocket thrust, if the weight flow rate through the nozzles is known. The higher the thrust, the higher is the spaceship's change in velocity. Second, it measures the engine efficiency. If two different rocket engines have different specific-impulse values, the engine with the higher specific impulse is more efficient because it produces more thrust for the same propellant mass. Table 2.1 shows specific impulses for different spaceships' engines. Chemical propulsion has the lowest specific impulse.

Electrical propulsion uses electrical energy to change the spaceship's velocity. It works by electrically expelling the propellant at high speed. Electric propulsion has a specific impulse of around 3,000 seconds. Due to limited electric power, the thrust is much weaker compared to chemical rockets, but since electric propulsion can offer a small thrust for a long time, it can achieve high speed. So, they are optimal for robotic exploration of the solar system. Electric propulsion, already tested successfully on many robotic missions, is a mature technology.

Nuclear propulsion produces high effective exhaust velocity, so it reaches higher specific impulses. It can overcome chemical rockets' specific impulse limitations because the energy source and the propellant

Table 2.1 Specific impulses in seconds

Type of spaceship engine	Specific impulse
Solid fuel	250
Liquid fuel	450
Ion	3,000
Nuclear fission	800–10,000
VASIMIR plasma	1,000–30,000
Nuclear fusion	2,500–200,000
Nuclear pulse	10,000–1 million
Matter–antimatter	1 million–10 millions

are independent of each other. Compared with chemical and electrical rockets, nuclear propulsion is more reliable and flexible for long-distance missions. It can achieve space missions at lower cost. In a nutshell, the reason for these advantages is that it can get "more miles per gallon" than chemical and electrical rockets.

Research on nuclear fission propulsion is well advanced in the United States, Russia, and Europe. In the United States, research on nuclear fission propulsion began in 1961 with the NERVA programme (Nuclear Engine for Rocket Vehicle Applications). During testing in the laboratory, the NERVA nuclear engine reached a specific impulse of 1,850 seconds. However, due to the general lack of interest in human space exploration at the end of Apollo project, NASA cancelled the programme in 1972. The Soviet Union had its own nuclear propulsion programme. The Russians tested the nuclear rocket RD-0410 near the test site at Semipalatinsk and proposed using it in the anticipated Kurchatov 1994 manned Mars mission. However, budget constraints and the lack of interest by the Soviet authorities for interplanetary missions put a stop to this nuclear propulsion research.

After 1972, research on nuclear fission propulsion continued in the United States and Europe. Research and technology efforts at NASA focused on nuclear thermal rockets and on electric nuclear propulsion.[18] Nuclear thermal rockets build on technologies developed under the NERVA programme. The main drawback of nuclear fission propulsion is the low level of technological readiness. Nuclear thermal propulsion has not such a low level of technological readiness. The tests performed in the 1970s on prototype nuclear thermal thrusters enables the development of safe, reliable thrusters in a reasonable time and at reasonable cost. The problem of the technological readiness of nuclear electric propulsion may be split into two parts: design of the power plant and design of the thrusters. In the former, the technology readiness has advanced; in the latter, electric thrusters are a mature technology on a small scale. Research aimed at scaling them up and improving their performance is under way.

[18] In a nuclear thermal rocket, a working fluid is heated to a high temperature in a nuclear reactor. The heated fuel then expands through a rocket nozzle to create thrust. In a nuclear electric rocket, nuclear thermal energy is changed into electrical energy that is used to power one of the electrical propulsion technologies.

More advanced research on nuclear-fission propulsion is being conducted in the United States and Europe. We shall mention two such propulsion engines here. The first engine, developed in the United States, is called the fission-fragment engine,[19] which directly harnesses hot nuclear-fission products for thrust as opposed to using a separate fluid as a working mass. Professor Carlo Rubbia and his colleagues advanced a second proposal, incorporated in Project 242.[20] This propulsion engine works on the principle of direct transfer of energy from the fission fragments to a gas to generate thrust. These two most advanced engines can reach a specific impulse of 30,000 seconds and more. In technical terms, they combine the thrust and specific impulse that must be maximised to move a large mass at high speeds.[21] These more advanced propulsion engines are optimal for human exploration of the near-Earth planets, such as Mars, because they reduce travel times and propellant mass. Although these engines are at the conceptual level, they are starting points for building nuclear-powered spaceships for human exploration to Mars.

Rockets that harness the power of nuclear fusion offer the next big leap in the quest to explore vast swathes of the solar system. A nuclear fusion rocket has the highest specific impulse with the exception of matter–antimatter engines. Such a rocket would provide efficient and long-term acceleration in space without the need to carry a large fuel supply and would slash travel times. NASA has pursued research into this kind of propulsion. It has funded several early stage fusion ideas via a programme called NIAC (NASA Institute for Advanced Concepts). NASA's Glenn Research Center has proposed a small-ratio spherical torus reactor for its Discovery II conceptual vehicle design.[22] The hydrogen is heated by the fusion plasma debris to increase thrust. The exhaust

[19] The Idaho National Engineering Laboratory and Lawrence Livermore National Laboratory developed this engine. The fuel is placed into several very thin carbon bundles, each one sub-critical. Bundles are collected and arranged as spokes on a wheel. Several wheels are stacked on a common shaft to produce a single large cylinder. The entire cylinder is rotated so that a few bundles are always in a reactor core. The surrounding fuel makes the bundles go critical.

[20] See Bignami G. et al. (2011).

[21] A spaceship is propelled forward by a thrust force equal to the mass flow rate multiplied by the exhaust particles' speed relative to the spaceship. It can thus achieve a final velocity that is higher than the velocity of the particles in its exhaust jet.

[22] See Craig H. W. (2005).

velocity is estimated at 348/463 kilometres per second, or 1/1,000 of the speed of light. It could deliver a 150–200 tonne payload to Jupiter in four months (one way). It uses 861 tonnes of hydrogen, plus 11 tonnes of deuterium.

The main alternative to magnetic confinement is inertial-confinement fusion (ICF).[23] A small pellet of fusion fuel is ignited by an electron beam or a laser. To produce direct thrust, a magnetic field will form the pusher plate. A new approach, the magnetic-target fusion (MTF),[24] combines the best features of the magnetic confinement and inertial confinement fusion approaches. As in the magnetic approach, fusion fuel is confined at low density by magnetic fields while heated into plasma. Unlike the inertial confinement approach, fusion is started by rapidly squeezing the target and so increasing fuel density and temperature. Fusion rockets will not be powering a spacecraft soon, since researchers have not yet developed fusion reactors that generate more energy than they consume. But studying how to make fusion work for spaceships helps chip away at this problem, perhaps bringing to reality a technology famous for always being "30 years away."

The large mass of fusion reactors is the main disadvantage for fusion powered spaceships. Recently, Lockheed Martin's research unit announced the intention to build and test a compact fusion reactor in less than a year.[25] They expect to complete the prototype in five to ten years, by using magnetic mirror confinement to contain the plasma in which fusion occurs. One of the project's innovations is using superconducting magnets that produce strong magnetic fields with less energy than conventional magnets. They plan to replace microwave emitters to heat the plasma with neutral beam injection. Through injection, electrically neutral deuterium atoms transfer their energy to the plasma. The high beta configuration allows a compact fusion reactor design. According to

[23] In the 1980s, Lawrence Livermore National Laboratory and NASA studied an ICF powered "Vehicle for Interplanetary Transport Applications" (VISTA). The conical VISTA spaceship could deliver a 100 tonne payload to Mars orbit and return to Earth in 130 days. VISTA needs 41 tonnes of deuterium/tritium plus 4,124 tonnes of hydrogen for this trip. The exhaust velocity was estimated at 157 km/s. The estimated specific impulse is around 70,000.

[24] The NASA Human Outer Planets Exploration (HOPE) group has investigated a manned MTF propulsion spacecraft.

[25] See http://www.lockheedmartin.com/us/products/compact-fusion.html.

the research director, the reactor's architecture reduces its mass by a factor of ten with respect to classical fusion reactors. If research were to be successful and costs contained, that would pave the way not only to produce electricity but to a practical employment on spaceships. The good news is that, if this technology is successful, we have now to wait only five to ten years.

2.3.4 The Birth of the Space Economy

Powerful rockets were instrumental in the birth of the space economy. In October 1957, two physicists at Johns Hopkins University decided to intercept microwaves coming from the Sputnik. They used a 20 MHz receiver to capture microwaves and transformed them into sound through an amplifier and calculated the satellite's speed by using the Doppler effect. This is possible due to the satellite first approaching the stationary receiver and then receding at a fixed frequency. By analysing the Doppler shift slope, they calculated the satellite's position. As in any university network, the news spread. The laboratory's deputy director asked if, knowing the satellite orbit and speed, an object's position on the ground could be determined. The following week, they delivered theoretical calculations on this problem. But they did not know that the United States Navy was already doing this so as to calculate the position of its submarines and enable continuous navigation. Five years later, the United States Navy had five satellites in orbit performing the job. Their research was the forerunner of the Global Positioning System (GPS).

The Soviet Union's success with Sputnik spurred the US Department of Defense to launch a research and development programme for long-distance satellite communications at 26 MHz. The department was interested in satellites for communications between their various units stationed around the world and contracted AT&T, Space Technology Laboratories, and Hughes Aircraft to do the job. Following this research, they put the satellite INTELSAT 1 into a geostationary orbit in 1965. It is important to note that the private companies took part in these activities through government contracts. As a result, public spending paid for the research and development, and the launches of military satellites reduced the business risk for future private activities. In 1968, the companies taking part in the Department of Defense programme, launched the first civilian long distance telecommunications satellite.

During these early years, the United States and Russia dominated satellite communications, though radical changes took place later globally. The "Open Sky" policy introduced by Nixon in the 1970s deregulated the American market allowing entry to many companies. Globally, an increasing number of countries entered the market. By 2011, more than 50 countries had launched their own satellites, and seven countries had operational launchers. China and India rose as established space powers alongside the re-emergent Russian Federation, which has launched more rockets than any other country every year since 2006.

Asian countries led by China are outdistancing Europe in terms of the number of launches and satellites sent into orbit. Geostationary orbit (GEO) satellites are the backbone of the commercial satellite communications industry. Radio waves travel in straight lines at the microwave frequencies used for wideband communications, so a repeater is needed to send signals over long distances. Repeaters can be located on satellites and can link places on the Earth that are thousands of kilometres apart, and large GEO satellites can serve one-third of the Earth's surface.

Satellite technology has now been extended to various services and applications, including products and services of information technology, television, GPS systems, and mobile phones. In 2011, the turnover of the satellite industry amounted to $177 billion (Fig. 2.3).

Direct Broadcast Satellite (DBS) television was developed in the United States to overcome "blind spots" in the coverage area of traditional terrestrial systems. Today, a DBS can send digital video signals from GEO directly to individual subscribers via small receiving dishes. There are other services that are focused on specific applications, including radio-navigation satellite and meteorological satellite services. Satellite technology's use in navigation, communications, meteorology, and Earth observation has given rise to a growing stream of applications, including air traffic control, transport, natural-resource management, agriculture, environmental and climate-change monitoring, and entertainment. They

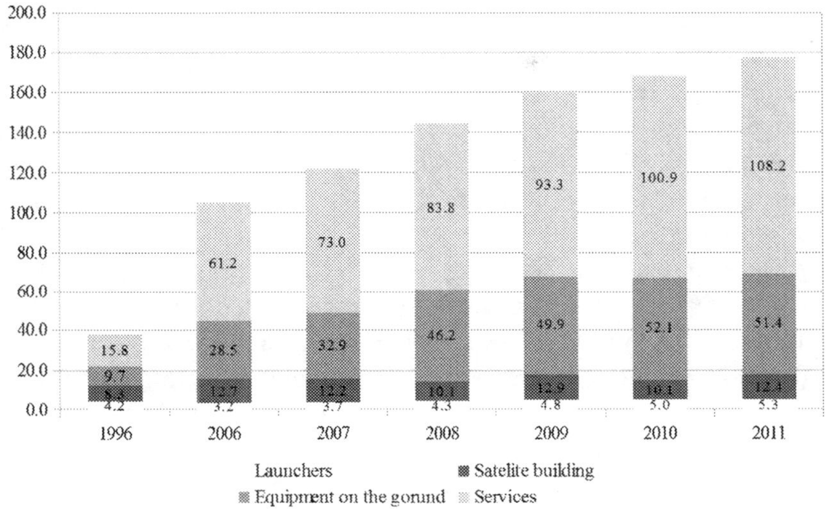

Fig. 2.3 Turnover of the satellite industry (*Source*: *The Space Economy at a Glance* (OECD 2011))

are creating new downstream uses and new markets. So space can be seen as an important potential source of economic growth, social well-being, and sustainable development.

The growth in the turnover of the satellite industry has remained strong despite the adverse economic environment: in 2013 turnover reached \$239 billion, a 35% increase over two years. Demand has remained particularly strong in emerging satellite markets, which include Latin America, Africa, Central Europe, and large parts of Asia. The services segment is the largest in the satellite industry and has witnessed the highest growth. Over the next ten years, the growing proliferation of digital content will fuel the demand for satellite infrastructure globally. The fast growth of satellites for communication and broadcasting is favoured by the newly competitive providers of video services, which offer coverage in rural areas with diverse service offerings and extensive sports programming.

Satellite technology's impact on the world economy is significant and can be divided into three categories: direct effects as measured by revenues generated by satellite companies; welfare effects on consumers as measured by benefits to consumers beyond the value they have paid for these services; and economic effects, arising from efficiencies generated by these services. The welfare effects include, among others, shorter automobile travel time and fuel savings generated by more efficient driving via GPS. These effects are estimated to be worth around $20 billion a year.

Economic effects are those leading to increased productivity and higher potential output and can be observed in the logistics sector. Cost savings generated by GPS services through fuel savings are estimated at around $10 billion per year. A recent study shows that meteorological-satellite services help to reduce water use and obtain higher yields per hectare of cultivated land, with an economic impact estimated at $40–110 per hectare. Earth-observation services also have major economic impacts and work through processes such as the search for new sources of raw materials in inaccessible places and the monitoring of extreme events. Information (via the internet and mobile phones) creates efficiency and economic opportunities by reaching areas of the world that are difficult otherwise to reach.

An international regulatory framework facilitates the satellite industry's development, without which private companies and financial institutions would not be able to invest large amounts of capital because legal and political difficulties would undermine their activities. The general principles of this framework are in the Outer Space Treaty of the United Nations, which provides slots for GEOs that are assigned by nations-states and not directly to private companies. As a result, the slots fall under the jurisdiction of the International Telecommunications Union (ITU) and governmental regulative agencies. The ITU assigns both orbital and electromagnetic spectrum positions. The rights and obligations are binding on member states. The ITU applies the "first-in-time, first-in-right" rule for orbital allocations. States carry on their registers the objects they have launched into space. According to Article VII of the Outer Space Treaty, a state retains jurisdiction and control over the object and the personnel involved in its operation and maintenance.

Table 2.2 Costs of the heavy launchers

Vehicles	Ariane 5G	Long March 3B	Proton	Atlas V	Falcon Heavy
Country	Europe	China	Russia	United States	United States
Load capacity (kg)	6,800	5,200	4,630	3,750	6,400
Launch costs ($millions)	165	60	85	100	85
Costs ($/kg)	24,265	11,539	18,359	26,667	13,280

Source: FAA Office of Commercial Space Transportation and official SpaceX website (www.spacex.com).

2.3.4.1 The Launchers

Immediately after the shuttles' retirement, the United States and other Western nations depended on Russian rockets for servicing the ISS. Since then, NASA transitioned from a government owned and operated delivery system to the ISS to a privately owned and operated one with multiple competitors. Two major policy changes by NASA favoured this transition. The first one was the opening of the launching sector to new companies, stimulating competition. NASA began a multiphase space technology development programme called the "Commercial Crew Transportation Capability" (CCDev), which is funded by the US government and administered by NASA. CCDev's main goal is to foster research and development into human space flight concepts and technologies in private companies. The programme has stimulated the development of privately operated crew vehicles. The result of these policies has been the successful development of two new launch vehicles (SpaceX's Falcon 9 and Orbital's Antares) and two new ISS cargo delivery spacecraft (Dragon and Cygnus). NASA then signed commercial, fixed-price contracts with selected private companies to deliver cargo to the ISS and is in the process of repeating this exercise for commercial crew transportation capability. The target is for the United States to become again independent in launching human space missions into low earth orbit (LEO).

The second major policy change concerns the way NASA makes contracts with outside companies. Back in the early days of the space race, people had no idea how much rockets would cost. So NASA used what it called "cost-plus" contracts that paid companies the cost of making a rocket, plus either a fixed fee or a percentage. Profit-wise, this was a pretty sweet deal for companies like Lockheed Martin and Boeing, but they didn't get to own the rockets or technology they built; NASA did. The agency also had a lot more say in determining how the spacecraft were designed and what tests they needed to pass.

Nowadays, companies have pretty much figured out how much to charge NASA to make rockets, so the agency uses "fixed-price" contracts. The companies quote NASA a price, and the space agency doesn't have to pay them more if they end up exceeding the budget. With the fixed-price contracts, NASA has less ability to dictate the design or the way the companies work as long as they meet its requirements. Now, the companies own the design, they actually own the vehicles, and NASA in a sense is just hiring rides to the ISS. And because the companies now get to hold onto their intellectual property, they have the opportunity to commercialize their rockets and eventually take civilians on-board.

These start-ups have made progress in the design and development of efficient and cheaper launch vehicles. Their cost advantages are not derived from advanced engines, but rather from their systems design and vertically integrated production, efficient supply chains, and efficiency in operations. Table 2.2 shows the costs of heavy launchers of the main spacefaring nations, including a start-up, SpaceX.

Experiments are ongoing in the United States to reuse the primary stage of a rocket. Blue Origin has succeeded in launching a rocket into space and landing it back at the company's Texas space port. SpaceX has recently succeeded in landing an F9R, which is basically a Falcon 9 first-stage rocket that has landing legs. An earlier test flight involving an ocean splashdown was successful. This proved that the Space X Falcon 9 can re-enter Earth's atmosphere and make a touch down at near zero velocity. While the landings were on target, the fins, which balance the rocket during its descent, ran out of hydraulic fluid just before the attempted touch down. The rocket crashed. A new test carried out in December 2015 succeeded in landing the first stage of the rocket on land. In April 2016, SpaceX finally achieved a sea landing when the company's first-stage booster glided smoothly to a hover before landing on a floating drone ship.

Table 2.3 Private space initiatives

Companies	Mission
Commercial lunar development	
Golden Spike Company	Golden Spike Company is planning affordable priced lunar orbital and surface expeditions.
Shackelton Energy	Shackleton intends to undertake lunar prospecting. They plan to establish a network of "refuelling service stations" in low Earth orbit and on the Moon to process and provide fuel and consumables for commercial and governmental customers.
Moon Express	Moon Express plans to offer commercial lunar robotic transportation and data services with a long-term goal of mining the Moon.
Excalibur Almaz	Excalibur Almaz plans to implement lunar tourism, plus wider space services, and space-flight technology.
Asteroid development companies	
Planetary Resources	Planetary Resources' short-term mission is to develop low-cost robotic spacecraft to explore resource-rich asteroids. The long term mission is to develop the most efficient capabilities to deliver these resources directly to both space-based and terrestrial customers.
Deep Space Industries	Their mission is asteroid mining and plan to offer space-based refuelling, power, asteroid mining, and space manufacturing.

The recent successes of Blue Origin and SpaceX in recovery tests of the rocket's primary stage may lead to innovations which would further reduce launch costs.[26] Previously, the primary stage of the rockets—responsible for the initial thrust through the Earth's atmosphere—simply detached after launch and fell back into the ocean as waste. By developing methods for reusing the primary stage, these companies stand to save tens of millions of dollars per flight. Depending on the amount of work required for the maintenance of the primary stage after landing, costs per kilogram of payload could be reduced by 30–40 %. Moreover, SpaceX is working to build a rocket on which both the first and second stages are reusable. If fully reusable, this could reduce the costs per kilogram of payload to around $1,000.

Competition is producing pressure on other aerospace companies in the United States and Europe to review their technologies and reduce

[26] SpaceX claims that the primary stage's costs represent around two-thirds of the total cost of the rocket. SpaceX has said that if they are successful in developing the reusable technology, launch prices of around 5 to 7 million US$ for a reusable Falcon 9 are possible.

Table 2.4 List of space investing billionaires

Rank	Name	Age	Net worth ($ billions)	Source	Space investment
19	Jeff Bezos	49	25.2	Amazon	Blue Origin
20	Larry Page	40	22.8	Google	Google Lunar X Price
21	Sergey Brin	40	23.0	Google	Planetary Resources
53	Paul Allen	50	15.0	Microsoft	Spaceship One
138	Eric Schmidt	68	8.2	Google	Planetary Resources
272	Richard Branson	53	4.6	Virgin Group	Virgin Galactic
527	Elon Musk	42	2.7	PayPal	SpaceX
831	Guy Laliberte	53	1.8	Cirque du Soleil	Visitor to ISS
922	Rma Shriram	56	1.7	Google	Planetary Resources
1,051	Ross Perot Jr	54	1.4	Oil & Gas	Planetary Resources
			106.4		

costs. Boeing has allied itself with Blue Origin[27] to develop an engine for its Atlas 5 rocket that will replace the Russian engine currently being used. Lockheed Martin has unveiled a new space-flight architecture to take cargo to the ISS and help in interplanetary flights.[28] Europe has responded to the competition from the United States by developing a new-generation launcher: Ariane 6. This follows the decision taken at the ESA Council meeting at the ministerial level in December 2014 to maintain Europe's leadership in the fast-changing commercial launch service market while responding to the needs of European institutional missions.

[27] The rocket developed by Blue Origin is powered by a mixture of oxygen and natural gas. Blue Origin is developing a technology to recover the first stage of the rocket, which should reduce launching costs.

[28] It features a space tug called Jupiter and a space module called Exoliner. The Jupiter module is based on the NASA–Lockheed-built MAVEN (Mars Atmosphere and Volatile Evolution Spacecraft), which entered Mars orbit in 2014. According to Lockheed Martin, Exoliner can be converted into a human habitat from its present cargo configuration. The two vessels will be used, in concert with NASA's Orion capsule, to place astronauts into Moon orbit. Such a mission will help us to learn how to live in deep space outside the Earth's protective magnetic field.

The US launch market may have two competitive super-heavy launch vehicles available by the 2020s to launch payloads of 100 tonnes or more into low-Earth orbit. The US government is currently developing the Space Launch System (SLS), a heavy-lift launch vehicle for very large payloads of 70 to 130 tonnes. On the commercial side, SpaceX is privately developing a super-heavy lift launch vehicle, which is being designed to lift a 100 tonne payload into Earth orbit. SpaceX has played down the competitive aspect with SLS. However, if the company makes progress with this vehicle in the coming years, it is almost unavoidable that America's two HLVs will attract comparisons and a healthy debate.

The United States is thus regaining its competitive position vis-à-vis Russia and China. On the other hand, Europe has remained tied to the old model, characterized by state procurement contracts and a monopolistic launching industry. At the moment, the only company in Europe producing medium to heavy rockets (Series Ariane) is Arianespace, which does so for the launch of commercial satellites. It has a close relationship with ESA for the launch of the latter's satellites and scientific missions. Stiff competition on price has induced seven European satellite operator companies, including the four largest in the world by annual revenue, to request that ESA find immediate ways to reduce Ariane 5 rocket launch costs and, in the longer term, make the next-generation Ariane 6 vehicle more attractive for telecommunications satellites. However, if the new American start-ups succeed in reusing the primary stage of their rocket, it is doubtful that the next-generation Ariane 6 will be able to compete.[29]

2.3.4.2 New Private Space Initiatives

Within the last decade, several private initiatives have surfaced, promoting space exploration and development. A partial list of these companies is shown in Table 2.3.

---

[29] Proposed launch prices are €69 million for Ariane 6.2, which might compete with Falcon 9. However, if SpaceX succeed in reusing the primary stage of Falcon 9, its costs could decline by 30–40%, making Ariane 6.2 uncompetitive.

These companies are seeking the best path to commercial success and aggressively leading the way beyond Low Earth Orbit LEO. One short-term strategy would produce fuel, oxygen, and water for fuelling spaceships, existing satellites, and life-support systems. A medium-term strategy is centred on leveraging space minerals into a logical configuration of complexity and infrastructure needs. As humanity moves towards the asteroids, the use of minerals found throughout the solar system will be instrumental to their success. The question is not "how" to leverage space-mineral resources, but "how best" to leverage them. Among their exploitation will be to use space minerals for survival, habitats, and space infrastructures during humanity's expansion into space, and when profitable for trade with Earth.

Investment in private space initiatives by members of the Forbes Billionaires is becoming increasingly fashionable. A list of six space-investing billionaires in 2011 grew to ten in 2013. The joint net worth of these individuals is over $106 billion as shown in Table 2.4. Today, other high net-worth individuals investing in space include Robert Bigelow (Bigelow Aerospace), Charles Simonyi (Planetary Resources), Anousheh Ansari (X-Prize), Dennis Tito (Inspiration Mars), Naveen Jain (Moon Express), Barney Pell (Moon Express), Tom Pickens (SpaceHab), and John Carmack (Armadillo Aerospace).

2.4 Lessons from History

A main lesson from history shows that the geopolitical conditions prevailing in the 1960s were unique, and most likely they will not reappear. Strong governmental commitment to human space exploration is gone. But, in the last decade, wealthy individuals and private companies have demonstrated a commitment to space exploration and development. Cooperation between space agencies and private companies could be a way to move human space exploration forward, but such indispensable cooperation requires an understanding of the economic and political benefits to be derived from these missions. Only in this way can an enduring consensus be achieved among policy makers and the public concerning programmes for long-term space exploration.

A second lesson strongly highlights the need for space infrastructure to enable and sustain interplanetary human exploration. In the short to medium term, these infrastructures are: (i) a space station serving as an assembly, staging, and training base; and (ii) a fleet of low-cost vehicles to bring cargo and crews to and from the space station. In the longer term, a permanent human outpost on Mars and nuclear-powered spaceships are needed. This outpost will serve the spaceships travelling between Earth and Mars, and beyond. A third lesson identifies the need of international cooperation among national space agencies due to the large costs of building a space station, and also the need among nations to enact international legislation on space resource uses.

The following three chapters analyse a scenario of potential missions comprising human space exploration and colonisation: a "planet hopping" approach in the sequence of Moon, near Earth objects, the moons of Mars, and Mars itself, and in the longer term the extra solar planets. The purpose of the initial human outpost is not to be there and look cool, nor is it to unfurl flags and take pretty pictures, nor is it the holy grail of science—although we will get all of those things. The main purpose is to exploit space resources such as minerals, water, and volatiles. As humanity moves towards other planets in the solar system and beyond, the use of space resources will be instrumental to success. The final chapter analyses the risks and uncertainties in this scenario.

References

Bignami G et al. Project 242: fission fragments direct heating for space propulsion – program, synthesis and applications to space exploration. Astro Acta. 2011a;82:153–8.

Chase Econometric Associates. The economic impact of NASA R&D spending: preliminary executive summary. NASA Technical Note 2741; 1975 Apr.

Craig HW. NASA TM-2005-213559, 2005. http://ntrs.nasa.gov/archive/nasa/casi.ntrs.nasa.gov/20050160960_2005161052.pdf

Hertzfeld HR. The impact of NASA research and development expenditures on technological innovation and the economy; 1998.

Launius RD. NASA: a history of the US civil space program. Malabar: Krieger Publishing Company; 1994.

Minovitch M. A method for determining interplanetary free-fall reconnaissance trajectories. Jet Propulsion Laboratory Technical Memos (TM-312-130); 1961 Aug. p. 38–44.

OECD. The space economy at a glance. Paris: OECD Publishing; 2014.

Shklovsky IS. Five billion vodka bottles to the moon: tales of a Soviet scientist. New York: W.W. Norton & Company; 1991.

Shoemaker et al. Earth approaching asteroids as targets for exploration. In Asteroids Exploration Assessment. NASA Conference Publications; 1978 Jan. p. 245–56.

Tsiolkovsky K. Exploration of the universe with reaction machines: exploring the unknown. The NASA History Series, NASA SP 4407. Washington DC; 1995.

von Braun W. The Mars project. Urbana: University of Illinois Press; 1952.

Wilford J. We reach the moon; the New York times story of man's greatest adventure. New York: Bantam Paperbacks; 1969.

3

Human Space Exploration in the "Deep Space Proving Grounds"

Freeman Dyson once said "there is nothing so great or so crazy that a technological society cannot be moved to do, provided that it is physically possible." He is one of the few surviving giants among visionaries from the space race of the 1960s. In his book[1] *Disturbing the Universe*, he presents heretical ideas, such as space colonisation and the exploitation of natural resources in space. He examines two scenarios. The first one explores the feasibility of a habitat around the Sun, which uses its energy to sustain life. Public money would finance this project. The second scenario analyses the activities of small groups of settlers, who use the resources of asteroids to support their operations. His preferences lie with the latter scenario:

> I did some historical research on the costs of the Mayflower journey, and of the migration of Mormons in Utah, and I think that we can go into space with a much lower cost. A cost of about 40,000 US$ per person (in 1978 dollars or 143,254 in 2013 dollars) would be the goal to be achieved; in terms of real wages, it would make it comparable to the colonization of America. Unless the cost is brought down to that level, these projects are

[1] See Dyson F. J. (1979).

G. Bignami, A. Sommariva, *The Future of Human Space Exploration*,
DOI 10.1057/978-1-137-52658-8_3

not very interesting to me, because otherwise it would be a luxury that only governments could afford.

These ideas, which appeared then to be only science fiction, are now the goals of space agencies and private companies. Two general approaches to space mining are debated among private companies: lunar and asteroid mining. The relative merits of lunar versus asteroid have not yet been decisively determined. But the prevailing opinions favour asteroid mining.

3.1 The Exploitation of the Moon's Mineral Resources

The Moon is the near-Earth object most extensively explored. Its physical exploration began in 1959 with the landing of the Soviet Luna 2.[2] From the mid-1960s to the mid-1970s, there were 65 Moon landings, both manned and robotic. Perhaps most celebrated, the seven Apollo missions brought back to Earth 380 kg of samples from the lunar surface. After 1976, lunar exploration halted and only resumed in 1990 when Japan landed the Hiten spacecraft. ESA sent a small, low-cost lunar orbital probe in 2003. Its primary goal was to take three-dimensional X-ray and infra-red images of the lunar surface. In 2009 NASA sent the Lunar Reconnaissance Orbiter also to collect extensive imagery of the surface. It carried the Lunar Crater Observation and Sensing Satellite, which looked for the possible existence of water. In 2007 the People's Republic of China began a programme of Moon exploration to investigate the feasibility of lunar mining, specifically looking for the isotope helium-3 for use as an energy source on Earth.

The Moon's exploration and analysis of the samples brought back to Earth show the Moon's geological composition to be similar to Earth's (Table 3.1).

Absent of any substantial surprises, the Moon is lacking minerals marketable on Earth. The only valuable minerals are helium-3 and rare

[2] Luna 2 was a space probe launched by the Soviet Union which impacted onto the Moon's surface in September 1959.

Table 3.1 Abundance of chemicals and metals on the Moon

Components	A 11	A 12	A 14	A 15	A 16	A 17	Luna 16	Luna 20
SiO_2	42.47 %	46.17 %	48.08 %	46.20 %	45.09 %	39.87 %	43.96 %	44.95 %
Al_2O_3	13.78 %	13.71 %	17.41 %	10.32 %	27.18 %	10.97 %	15.51 %	23.07 %
TiO_2	7.67 %	3.07 %	1.70 %	2.16 %	0.56 %	9.42 %	3.53 %	0.49 %
Cr_2O_3	0.30 %	0.35 %	0.22 %	0.53 %	0.11 %	0.46 %	0.29 %	0.15 %
FeO	15.76 %	15.41 %	10.36 %	19.75 %	5.18 %	17.53 %	16.41 %	7.35 %
MnO	0.21 %	0.22 %	0.14 %	0.25 %	0.07 %	0.24 %	0.21 %	0.11 %
MgO	8.17 %	9.91 %	9.47 %	11.29 %	5.84 %	9.62 %	8.79 %	9.26 %
CaO	12.12 %	10.55 %	10.79 %	9.74 %	15.79 %	10.62 %	12.07 %	14.07 %
P_2O_5	0.12 %	0.10 %	0.09 %	0.06 %	0.06 %	0.13 %	0.21 %	0.08 %
H	51 ppm	45 ppm	80 ppm	64 ppm	56 ppm	60 ppm		
He	60 ppm	10 ppm	8 ppm	8 ppm	6 ppm	36 ppm		
C	135 ppm	104 ppm	130 ppm	95 ppm	106.5 ppm	82 ppm		
N	119 ppm	84 ppm	92 ppm	80 ppm	89 ppm	60 ppm	134 ppm	107 ppm
Rare earth elements					903 ppm			

Source: Beauford R. (2011); G. Kramer et al. (2003); and http://www.permanent.com/lunar-geology-minerals.html

Note: A11–A17 are Apollo missions

earth elements. Helium-3 is a scarce element on Earth. Used in industry, it needs to be manufactured and is a product of the decay of tritium, which results from neutron bombardment of deuterium, lithium, boron, or nitrogen targets. The current industrial consumption of helium-3 is around 60,000 litres per year. The price has varied from a low of $100 per litre to $2,000 per litre in recent years. Commercial fusion reactors using helium-3 are not yet available, and even if they do become commercially viable, they will need tens of tonnes of helium-3 each year to produce a fraction of the world's power. Recoverable helium-3 resources on the Moon are estimated at around 75,000 tonnes, which could satisfy future demand for use in prospective nuclear-fusion reactors.

Currently, the dominant end uses for rare earth elements are in automobiles and petroleum refining catalysts, phosphors in colour television and flat-panel displays (mobile phones, portable DVDs, and laptops), permanent magnets and rechargeable batteries for hybrid and electric vehicles, and numerous medical devices. At present, worldwide reserves of rare earth elements are estimated at 110 million tonnes, but world demand is estimated at 136,000 tonnes, while mine production is about 134,000 tonnes. The difference is made up by above-ground stocks or inventories. World demand is projected to rise to at least 160,000 tons annually by 2016 according to the Industrial Minerals Company of Australia. Based on the current and prospective demand, current worldwide reserves should suffice for about 600 years.

China has a near monopoly in global production of rare earth elements. Although it produces around 97% of global supply, only 50% cent of world reserves are actually in China. Rare earth element mines in North America closed because of environmental contamination problems, but China appears to be willing to bear the environmental cost of extraction. Due to high domestic demand, the Chinese government cut export quotas in 2010 and 2012. This caused a rapid increase in prices. These high prices and the vulnerability of the supply chain have stimulated exploration in other parts of the world. Exploration efforts, economic evaluations, and the search for solutions to environmental problems continue in North America and other countries.

Among rare earth elements, neodymium is of particular interest. Neodymium magnets are the strongest permanent magnets known and appear in products such as microphones, professional loudspeakers, in-ear headphones, guitar and bass guitar pick-ups, and computer hard disks where low mass, small volume, or strong magnetic fields are required. Due to this high magnetic capacity per weight, neodymium is used in the electric motors of hybrid and electric automobiles, and in the electricity generators of some designs of commercial wind turbines (those with "permanent magnet" generators). For example, the electric motors in each Toyota Prius require one kilogram of neodymium. Given current reserves, there are no problems for supply to satisfy demand. However, it is difficult to predict technological breakthroughs. If electric cars come to dominate the auto market, previously powered by the internal combustion engine, demand for neodymium will skyrocket. Consequently, off world supply will become an even more interesting economic proposition.

The Moon has abundant mineral resources which could be traded within the space market. Water and volatiles can be used to supply oxygen for life support and propellant for spaceships. Iron, nickel, titanium, and other minerals can be used to build space infrastructure. But even here, minerals from asteroids would out-compete minerals from the Moon; their competitive advantage stems from the comparative requisite changes in velocity required by missions (Table 3.2). Velocity changes from low-Earth orbit (LEO) to the Moon are higher than the ones to near-Earth asteroids (NEAs). Similarly, velocity changes from the lunar surface to LEO are higher than the ones from NEAs to LEO. Thus less energy is needed for asteroid missions than for Moon missions, and hence the competitive advantage of NEAs for the supply of minerals to the space market.

Table 3.2 Change in velocity required for transfer of minerals to Earth orbit

Transfers	Change in velocity (km/s)
Earth surface to LEO	8.0
LEO to escape velocity	3.2
LEO to Moon landing	6.3
LEO to NEAs	4.0
Moon surface to LEO	2.4
NEAs to Earth transfer	1.1

Source: Lewis, J. S. (1996)

The Moon, however, is a good platform for assembling and launching spaceships on interplanetary exploratory missions. Lunar gravity is lower than Earth gravity, and lunar minerals can be used to produce spaceships' parts. All the above could justify a permanent lunar base, though the Moon does not constitute an optimal platform to launch spaceships on interplanetary missions. To escape the Moon's gravity requires energy. At Lagrangian points, lunar and Earth gravitational influences cancel, and spaceships do not need energy to break free from the Moon's gravity. Since the Moon has many mysteries that need unravelling and that are important for understanding the origins of the Earth and the entire solar system, scientific knowledge is another argument favouring a permanent lunar base. The lunar surface is also an ideal platform for astronomical observations. It is stable, with low gravity, and an absence of wind or other significant forces; so it's an ideal place for a telescope. The bulky pointing devices of orbiting telescopes or the large-support structures of terrestrial telescopes are not needed on the Moon, though, ultimately, it is doubtful whether scientific reasons alone would justify the large investment of public resources required here. The debate continues, and many people are questioning the rationale of a permanent lunar base, among them Buzz Aldrin,[3] the second astronaut to put his feet on the Moon. Although not opposed to sending astronauts back to the Moon, he favours manned missions to Mars.

3.2 The Near-Earth Asteroids

The prime targets for asteroid mining are near earth asteroids (NEAs), which have orbits bringing them close to Earth on a regular basis. Most NEAs are likely the result of collisions between asteroids in the main asteroid belt, triggered by Jupiter's gravitational influence. A few NEAs may be the spent nuclei of comets. They are grouped into three categories: (i) Amors, (ii) Apollos, and (iii) Atens. Amors cross the orbit of Mars, but do not reach Earth orbit, for example Eros. Apollos cross Earth's orbit with a period longer than one year, for example Geographos. Atens cross

---

[3] See Buzz Aldrin (2013).

Earth's orbit with a period shorter than one year, for example Ra-Shalom (Fig. 3.1).

In the last decade, analysis of asteroids' mineral composition has advanced via spectroscopic and dynamic studies of asteroids, and analysis of meteorites that have fallen to Earth. The majority of known asteroids fall into three categories. Type C (carboniferous) asteroids account for around 75 %, are dark with an albedo between 0.03 and 0.09, and contain large amounts of carbon molecules, metals, water, and volatiles. Type S (siliceous) asteroids account for around 17 %, are relatively bright with an albedo between 0.10 and 0.22, and are estimated to be composed of

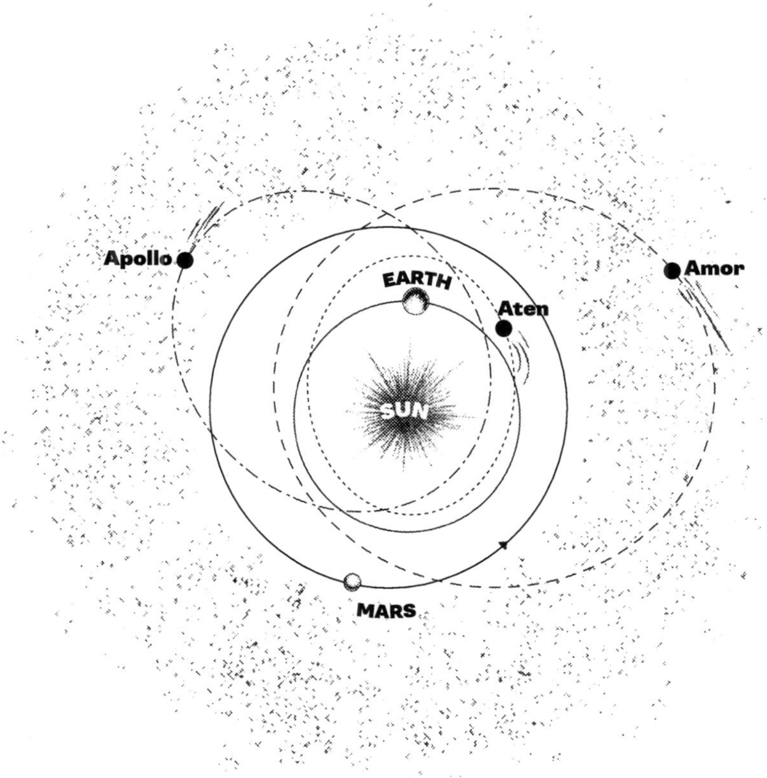

Fig. 3.1 Map of the asteroids

iron-nickel mixed with silicates. Type M (metallic) asteroids account for the rest of the known asteroids, are bright with an albedo between 0.10 and 0.18, and are composed predominantly by iron-nickel. Results from spectroscopic and dynamic studies are confirmed by analysis of meteorites that have fallen to Earth. Asteroids contain valuable elements such as the platinum group metals, rare earth elements, gold and germanium. Identified minerals include nickel, in the form of metallic nickel-iron, and cobalt (Table 3.3). Type C asteroids contain water and volatiles, valuable resources for in-space fuel and life-support systems.

The concentrations of Platinum Group Metals (PGMs) and precious metals vary with the type of asteroid, but are higher than in the Earth's crust. Most metals sank to the Earth's core when the planet was young, while con-

Table 3.3 Abundance of PGMs and precious metals in asteroids

Metals	Asteroid LL-chondrite	"Good" metallic asteroid	"Best" metallic asteroid	Earth's crust
Industrial elements				ppm
Cobalt	1.57 %	0.46–0.80 %	0.43–0.75 %	25
Nickel	34.3 %	5.6–18.0 %	5.4–16.5 %	120
Iron	63.7 %	81.0–94.0 %	82.0–94.0 %	55,000
Germanium	1,020	0.06–70	0.05–35	1.8
PGM	ppm	ppm	ppm	ppm
Rhenium	1.1	1.1	2.4	0.0004
Ruthenium	22.2	20.7	45.9	0.001
Rhodium	4.2	3.9	8.6	0.0002
Palladium	17.5	12.6	12	0.0006
Osmium	15.2	14.1	31.3	0.0001
Iridium	15.0	14.0	31.0	0.0003
Platinum	30.9	28.8	63.8	0.005
Precious metals				
Gold	4.3	0.16–0.70	0.06–0.6	0.001
Rare Earth Elements (REE)	0.4			0.2

Sources: Calculated from data given by Buchwald V. F. (1975); Malvin D. J. et al. (1984); Hoashi M. et al. (1993); and R. P. Mueller et al. (2010)

Notes: ppm (part per million) is a way of expressing very dilute concentrations of substances. Such measurements usually describe the concentrations of something in water or soil. One ppm is equivalent to 1 milligram of something per litre of water (mg/l) or 1 milligram of something per kilogram of soil (mg/kg)

centrations in the asteroids are higher because of low gravity. They reflect the original distribution of gases and dust from which the solar system originated. It is estimated that their raw materials are of excellent quality and the analysis of meteorites shows minerals in a free state. If confirmed, their minerals will need little machining during future mining operations.

3.3 What Will We Do in the Short Term: The Exploration of Asteroids

In the mining industry, estimates shown in Table 3.3 are referred as "mineral potentials." No mining company will invest unless the mineral potential is confirmed through further exploration. On Earth, providing geological data on regions by governments is seen as a public good. It is recognized as important to attract private companies and for the continued development of mineral resources. The same logic applies to asteroids: exploration is the first step, which can open the door to the exploitation of natural resources in space by private companies. So, by parallel thinking, technologies to reach and land on asteroids, and the identification and verification in situ of their composition, are public goods provided by space agencies.

Recently, NASA has approved the "Asteroid Redirect Mission" (ARM) whose goals are: (i) the identification, capture, and transport of an asteroid in a stable lunar orbit; and (ii) human missions to this asteroid while in lunar orbit. The corollary goals are testing new propulsion systems and the difficult manoeuvres of rendezvous with the asteroid, and analysing the effects on astronauts of long sojourns in space outside Earth's magnetic field. ARM would thus enable the assessment of the composition of mineral resources of the asteroid and the testing of methods and systems to carry out these operations. These are also valuable pieces of information for space mining companies which would reduce the risks of future operations on an industrial scale.

NEAs are good candidates for ARM. According to the International Astronomical Union, there are 10,337 known NEAs, 861 of them having diameters greater than one kilometre. Accessible asteroids are fewer. They

Fig. 3.2 Design of the spaceship (*Source*: Rick Sternbach/Keck Institute for Space Studies)

are defined by their orbital characteristics[4] (low inclination, eccentricity, rotation rate, and a distance from Earth of less than one AU), spectral class, and mass. They are dynamically more accessible than a round-trip mission to Mars, many (hundreds) require less energy than a round trip to the lunar surface, and dozens demand less energy than does a low orbit around the Moon. The sizes of asteroids range from a few metres to a few kilometres. Size is important for security reasons. If something were to go wrong, an asteroid, whose diameter is below 20 metres, would disintegrate if it enters the Earth's atmosphere.

NASA plans to launch a spaceship to recover an asteroid or a part of it in 2019, capturing it with an inflatable structure similar to a bag (see Fig. 3.2). Advanced solar-electric propulsion (SEP), called an ion

[4] Inclination and eccentricity determine changes in velocity from LEO to the asteroid and from the asteroid to LEO. The rate of rotation of an asteroid determines its accessibility. A slower rate of rotation is important for the possibility to anchor the spaceship to the asteroid.

thruster, will propel the spacecraft and high-efficiency ring-shaped solar panels will provide the electrical power. NASA is considering two options for capturing an asteroid. The first consists of capturing a small asteroid in its natural orbit; the second is to retrieve part of a larger asteroid. NASA has three small asteroid candidates:[5] 2009 BD, 2011 MD, and 2013 EC20, and three large-size candidates with small boulders:[6] Itokawa, Bennu, and 2008 EV5. NASA is planning reconnaissance missions to these asteroids before the capturing mission.

Once the asteroid is brought into lunar orbit, a manned mission will leave to explore it during the first half of the 2020s. The astronauts will use an Orion capsule launched by the Space Launch System[7] (SLS) rocket. In a major effort of international cooperation, NASA has signed contracts with ESA to build the service module of the Orion capsule, which, constructed with an aluminium-lithium alloy, will serve as a power-generating system and electrical propulsion unit. It has several elements: (i) solar panels for generating electricity; (ii) radiators; (iii) reaction control thrusters and cylinders of oxygen and nitrogen for life support; (iv) one carbon dioxide scrubber; (v) waste-water recycling systems; and (vi) a main motor for propelling the capsule.

These missions will provide useful information for longer-term human missions to Mars, for which SEP will be an important part. Spaceships powered by SEP will transport large quantities of materials and fuel to Mars before the human mission. Compared with chemical propulsion, SEP will reduce the propellant mass by a factor of five to ten. The Orion mission will require a series of complex manoeuvres to rendezvous and dock with the asteroid in lunar orbit. These manoeuvres are similar to those needed for insertion into, and departure from, Mars orbit.

NASA has not yet decided where to park the asteroid. If it chooses a Moon–Earth Lagrangian point, it could launch a habitable space module

[5] The diameter of 2009 BD is estimated to be around 4 metres, that of 2011 MD around 6 metres. These dimensions are derived from data provided by the Spitzer Space Telescope.

[6] The Japanese mission Hayabusa analysed Itokawa, which has boulders of around 3 metres. Both 2008 EV5 and Bennu were analysed by radar. Data show that they have small sized boulders.

[7] The Space Launch System (SLS) is a heavy launcher developed by NASA. Various updates with more powerful versions are foreseen. These updates will enable the SLS to carry astronauts and equipment to various destinations beyond LEO such as on a circumlunar trajectory and to an NEA.

with which the Orion capsule could dock. This module will form the space station's core and could be expanded later to store minerals, water, and volatiles extracted from the asteroid. At present, astronauts depend on Earth for supplies and operational support, but robotic and manned missions should help in developing ways to break these terrestrially dependent bonds. Water and volatiles supply the raw materials and fuel needed for the spaceships and obtaining these materials from space to use them there is comparable to the early American settlers' experience, who lived off the land and did not depend on supplies brought from Europe.

In Europe, ESA has been developing programmes for the exploration of asteroids and comets and which will add to our asteroid database. ESA's Rosetta probe reached Comet 67P/Churyumov-Gerasimenko in 2014. The probe experienced fly-by encounters with many asteroids, which were fleeting but added to our knowledge of them. ESA is also sponsoring Spaceguard, a worldwide asteroid-tracking project that aims to plot the orbits of those asteroids on possible collision courses with the Earth. So far, it has logged more than 300. Furthermore, asteroid explorations are likely to feature in the Aurora Programme, a joint ESA–NASA long-term programme for the manned exploration of the solar system.

Recently, an Italian researcher[8] put forward a proposal for the exploration of asteroid 3753 Cruithne,[9] which is a five-kilometre-wide near-Earth object which shares an orbit with the Earth. The Italian's plan is to send a small spaceship propelled by an electric-ion engine to rendezvous with Cruithne where it will then deploy two smaller CubeSats to examine the asteroid more closely. Cruithne's large orbital inclination presents a particular challenge for any spacecraft to reach, one that certainly tests the mettle of any proposed space mission, which is to go straight from Earth to it without having to be slingshot en route. This will save on transit time and complexity while offering a highly flexible launch window.

[8] See Pergola P. (2013).

[9] Duncan Waldron discovered Cruithne on 10 October 1986. It has a normal elliptic orbit around the Sun, but a periodic revolution almost equal to that of Earth. Because of this, Cruithne and Earth appear to "follow" each other in their paths around the Sun. This makes it a co-orbital object, which when viewed from Earth is described as a horseshoe. Its inclination to the ecliptic is 19.82° and has an albedo of 0.12, which makes it an S type asteroid. Gravity at its surface is estimated at 1.622 metres per square second.

The spacecraft is expected to weigh around 100 kilograms and will reach Cruithne in around 320 days. While interesting, the proposal is still a proof-of-concept and needs further elaboration to reach the stage of providing the content for a real mission.

This proposal should be given serious consideration for two reasons. First, Cruithne is of great scientific interest, its composition preserving its original chemical make-up—unaltered by high internal pressures and temperatures—which will tell us more about how the solar system formed. Second, a survey mission will help pave the way to robotic-landing missions, human exploration, and asteroid-mining endeavours. Cruithne could contain the largest deposit of space minerals near the Earth. For example, assuming an average density of 2.5 grams per cubic centimetre and around 31 ppm of platinum, the estimated mass of platinum is around 330,000 metric tonnes.

Private space companies are designing and planning exploratory missions to the asteroids. Planetary Resources[10] is designing orbital telescopes to explore asteroids and which are envisioned as the first step forward in the company's asteroid mining ambitions. Deep Space Industries is designing and planning low-cost reconnaissance missions to near Earth asteroids to analyse their key aspects and bring back samples. This will help to select an asteroid for future mining missions. One of the major benefits to private companies from the NASA initiative is the ability to leverage NASA's NEA databases.

Space agencies are likely to find synergies with private companies regarding reconnaissance missions to asteroids, which will liberate resources for the former's strategic missions, for example robotically deflecting the asteroid to enable human exploration in the lunar vicinity. Agreements could be reached with private space mining companies, which would carry out reconnaissance missions to targeted asteroids and share information with space agencies for a price, which would be a fraction of the costs incurred if they carried out the missions. Alternatively, they could exchange information for participation in the manned missions, which

[10] The dimensions of 2009 BD is estimated to be around 4 metres in equivalent diameter. The dimension of 2011 MD is around 6 metres. These dimensions are derived from data provided by the Spitzer Space Telescope.

would enable the testing of mining methods and equipment, and reduce the risks of future operations being on an industrial scale. Both solutions are Nash equilibria, where each player's strategy maximizes his or her expected utility against the other players' strategies. No player would be better off by unilaterally changing his or her strategy. An important step would be to set up an adequate forum to discuss and exchange information on programmes, affordability, and the timeframes of possible missions.

3.3.1 Mining an Asteroid in Lunar Orbit

There are two ways to mine an asteroid: in lunar orbit or in its natural orbit. In the first solution, technical limitations and safety considerations restrict the dimensions of targetable asteroids. The technology exists to move an asteroid with a maximum diameter of 10–20 metres. Safety considerations impose a limit of 20 metres, as asteroids with a diameter less than this will disintegrate in Earth's atmosphere without endangering the safety of the lifeforms and infrastructure on Earth, in the case where an asteroid went astray.

Table 3.4 reports the masses of type "LL-chondrite" asteroids, their mineral quantities, and their values (at current prices).

The Keck Institute estimated that the mission's total costs would be around \$2.6 billion; launch costs are estimated at around \$0.5 billion. Table 3.4 shows that sales of industrial metals and water and volatiles could make the project profitable. Prices of industrial metals and water and volatiles obtained in this way make it impractical to trade these products on the Earth market, but they would be tradable on the space market. Presently, the only existing space market is that in fuel supply to satellites and the International Space Station. Feeding this small market makes mining of a small asteroid in lunar orbit feasible. Moreover, the fast development of metallurgical 3D printing and reduced orbital launching costs make space-based solar power (SPS) economically workable. These eventualities will create a larger market for raw materials from space.

SPS collects solar power in space for use on Earth. The general idea of using solar irradiation in space as an energy source was present in

Table 3.4 Asteroid mass scaling, volumes, and values

Diameter (m)	Asteroid mass (tonnes)	Volume of PGMs (tonnes)	Value of PGMs ($ millions)	Volume of industrial metals (tonnes)	Value of industrial metals ($ millions)	Volume of water and volatiles (tonnes)	Value of water and volatiles ($ millions)
7	450	0.05	1.4	359	3,589	91.4	914.9
10	1,309	0.1	4.2	1,043	10,436	266.1	2,661.4
15	4,416	0.5	14.0	3,518	35,205	897.8	8,978.3

Note: The asteroid mass has been estimated assuming an average asteroid density of 2.5 g/cm^3

the first publications by Konstantin Tsiolkovsky (1886), and it has been actively researched since the early 1960s. The United States, Europe, and Japan[11] have carried out extensive studies on SPS. In recent years, it has attracted the attention of other countries, particularly China.[12] A careful review of these studies shows that these initiatives are driven by an increasing demand for energy and the wish to find clean alternatives to fossil and nuclear fuels. As an example, in the wake of the Fukushima disaster, Japan doubled its efforts to find a workable alternative to nuclear power. An updated SPS proposal from the Japan Aerospace Exploration Agency seeks to solve the island nation's energy woes. They have devised a road map describing a series of ground and orbital stations leading to a 1-gigawatt commercial SPS. As planned, it would have the same output as a typical nuclear power plant.

Recent studies[13] show that SPS could become competitive with non-renewable resources to produce electricity. The major obstacles to realising these projects are not technical, but economic and legal. A major stumbling block is the cost of launching. SPS components, including fuel, would have to be lifted into a low-Earth orbit and then transferred to a geostationary orbit by a reusable rocket. Metallurgical 3D printing in space using minerals from space and recent and expected reductions in launching costs may be the keys to solving this economic problem. But how can the cost problems be solved? The two heaviest components of a commercial SPS are the solar-panel array and the antenna for sending electric power to a receiving station on Earth. These two components have a combined mass of around 34,200 tonnes. The materials needed to build these could come from asteroids brought into lunar orbit. Manufacturing[14] them in space with materials found there will significantly cut the cost of lifting them into a geosynchronous orbit. Recent

---

[11] See J. C. Mankins (2012).

[12] The China Academy of Space Technology (CAST) submitted to the government a study of space solar power. CAST's study addresses the key components and defines a baseline or reference system that will allow an accurate determination of mass and cost in China. Later, the Ministry of Industry and Information Technology activated, approved, and funded a concept design project.

[13] See J. C. Mankins (2011).

[14] Research on 3D printing of solar photovoltaic panels is well advanced, particularly those that are manufactured using inks and dyes. Industrial 3D printing is well advanced in metal powders. Recent experiments at the International Space Station demonstrate that 3D printing functions in

and expected reductions in launching costs would also reduce the cost of lifting the remaining components of the SPS.

3.3.2 Mining an Asteroid in Its Natural Orbit

D. W. Cox[15] first discussed asteroid mining in their natural orbits. Later on, D. L. Kuck and M. J. Sonter[16] analysed the subject in depth. In this solution, the reference market for minerals such as PGMs and precious metals is the Earth. Asteroid mining is like any other mining investment project, more uncertain for sure, but conceptually similar. A firm will invest to maximise the surplus of expected revenues over expected costs. Each stream is discounted to present values. The revenues from a successful project depend upon the date of completion, the quality of the end product, and rivals' reactions to a new firm's entrance. The costs of the project depend upon the state of technology, the quality of the end product, and the speed of development.

Let us begin by specifying the cost side in more detail. To bring the mass of metals to the Earth market, a firm sustains a stream of costs over several time periods. Initially, the mining company incurs costs related to (i) identifying the targeted asteroid; and (ii) research and development of mining and refining equipment for their use in space. Large investment costs occur at the time of building and launching into orbit the equipment, power supply, the cargo container, and the spaceship. These costs depend on the expected mass to be returned to Earth orbit and on the asteroid's orbital characteristics. The mass returned to Earth orbit depends on the length of the mining season and on throughput. The length of the mining season varies and depends on the orbital characteristics of the asteroid. Aten asteroids have a quasi-circular orbit, hence longer mining seasons, while Apollo asteroids have elliptical orbits and shorter mining seasons. Low eccentricity targets result in longer mining seasons, while high eccentricity targets in short mining seasons.

zero gravity. Small 3D printers, capable of replicating themselves, can be lifted into orbit and there make parts of yet larger printers.

[15] See Cox, D. W. et al. (1964).

[16] See Kuck D. L. (1995); and Sonter, M. J. (1997).

Throughput is the ratio of the mass of regolith or rocks treated per day to the mass of the equipment. This depends on engineering choices concerning mining and processing equipment and the mining season's length: longer seasons imply less demanding specifications on mining and processing equipment, and lower costs. Apart from determining the season's length, the asteroid's orbital characteristics influence the spaceship's costs. Asteroids with low eccentricity and low inclination may need transferring with continuous thrust, those with high eccentricity and those with low eccentricity but high inclination require impulsive transfer. In the first case, electric propulsion, such as ion propulsion, is an option. In the second case, nuclear propulsion is the only choice; however, a nuclear-powered spaceship costs more than one with electric propulsion.

In summary, outlays are bell shaped, with the peak rate occurring at the time of the spaceship's construction and launch, after which and before return to Earth orbit, spending is consumed by operational costs, which are minor compared to investment costs. A final cost occurs when transferring metals from Earth orbit to the surface. But no unique pattern of expenditure exists. Any pattern depends on management decisions on the rate of spending during research and development. Management can spend at higher rates in the early period to bring the project's operations forward to an earlier date. When time is saved by increasing the rate of spending beyond a level sustaining economies of scale, total development costs increase. Research and development is a heuristic process and often involves significant uncertainties. Alternative technical approaches with finite success probabilities can be explored in series until success emerges, but this takes time. Overall expected time can be reduced by running technical approaches concurrently, but this increases the expected project costs because more approaches are run, some of which ultimately prove unnecessary. The relationship between expected research and development time and expected total development costs is thus convex to the time and cost coordinates. As time is compressed, total development costs rise at an increasing rate.

A firm will adjust the time allocated to the R & D programme and the choosing of an asteroid to maximise the surplus of discounted expected revenues over total expected costs. Expected revenues depend on how other companies in the market react to the presence of a new firm. The

Table 3.5 Characteristics of Asteroid 2004 MN4

Orbital type	Aten
Eccentricity	0.191
Inclination to the ecliptic	3.331
Perihelion in AU	0.746
Aphelion in AU	1.098
Semi-major axis in AU	0.922
Mass of the asteroid in billion tonnes	40
Orbital period in days	323

PGM market is an oligopolistic one and there are few PGM-producing regions worldwide and few mining companies. Russia and South Africa account for around 80 % of the global PGM supply. The growth of supply is constrained by political, infrastructure and cost issues in South Africa; declining PGM production in Russia; and a few new projects. South African production is challenged by deeper mines, power and water limitations, higher operating costs, geopolitical risks, and skilled labour shortages. Coupled with rising mining cash outlays, these constraints will limit the supply of PGMs.

Opportunities exist for new entrants due to expected high demand and supply constraints. Incumbency gives oligopolistic firms no privileged position if costs are fixed rather than sunk. Such firms cannot earn large profits because the threat of entry drives the incumbent firm's profit to the competitive level. But the new entrant ought to devise a sale strategy to maximise its profits in time. So no unambiguous solution exists and the outcome will depend on specific assumptions about other companies' reactions. The economic literature has analysed these problems and suggests that an equilibrium solution exists if the new entrant's annual supply is small and information perfect. This solution is often criticised as unrealistic because it is rare that firms select quantities simultaneously without the ability to know their competitors' choices. Heinrich von Stackelberg pointed out that results will be different if firms choose quantities sequentially.

Stackelberg games are classic examples of the bi-level optimisation problems often encountered in game theory and economics. These are complex problems with a hierarchical structure where one optimisation task is nested within the other. In a multi-period strategy with imperfect information, a sequential equilibrium prevails. Each company infers information on other companies' behaviour by their past actions and adjusts the offers to made for the next period. The adjustment will be incomplete in any single period, but over the longer run a final state of balance results. Here the off-world miner reaches agreements with a service company, such as Johnson Matthey in London, which stores the metals returned to Earth. As the service company sells the metals through a series of forward contracts, it acts as liquidity provider and sales agent.

3.3.3 A Hypothetical Case

Mining asteroid 2004 MN4[17] is the mission in this scenario. A team of scientists led by Thomas Müller[18] made a series of observations of 2004 MN4 in the far infra-red part of the spectrum. This study suggests that the asteroid is composed of a small mass of fine regolith and a larger volume of denser rocks and boulders. The orbital and other characteristics of the asteroid are shown in Table 3.5.

This mission assumes Hohmann transfer to rendezvous with the asteroid at perihelion and near-aphelion return trajectory to Earth. Table 3.6 reports the changes in velocity (Δvs) needed for rendezvous with the asteroid at perihelion and the return of the payload to LEO from near aphelion. These transfers are lowest-energy transfers.[19]

Total time from launch to return in LEO is roughly the sum of three orbital half-periods, namely the outbound transfer orbit, the asteroid's half-period mining season, and the payload-return transfer orbit. Total

[17] Roy A. Tucker, David J. Tholen, and Fabrizio Bernardi discovered this asteroid at the Kitt Peak National Observatory on 19 June 2004. It has an eccentric orbit, which takes it from 0.746 AU from the Sun to 1.0985 AU. This makes 2004 MN4 an Aten type asteroid. Estimates point to an equivalent diameter of 375 metres. It has an albedo of 0.23, which makes it an S type asteroid.

[18] See Muller T. G. (2014).

[19] In orbital mechanics, a Hohmann transfer orbit is an elliptical orbit used to transfer vehicles between two circular orbits of different radii in the same plane. Orbital manoeuvres to perform Hohmann transfers use two engine impulses, one to move the spacecraft into transfer orbit and a second to move away from it. With a Hohmann transfer, alignment of two celestial bodies in their orbits is crucial. The destination celestial body and the spaceship must arrive at the same time at the same point in their respective orbits around the Sun. The requirement for this alignment implies that missions to asteroids can take place only at specific launch windows.

Table 3.6 Changes in velocity

$\Delta v_{\text{departure, leo}}$ in km/s	3.47
Δv_{return} in km/s	2.46
Hyperbolic velocity at arrival in km/s	2.79
Mining season in years	0.4
Total transfer time from and to LEO in years	1.3
Total mission time in years	2.4

Table 3.7 Scenario

Mass returned into Earth orbit in tonnes	736
Of which PGMs (tonnes)	574
Period of sales, in years	11
Average quantity sold per year in tonnes	72.9
Of which PGMs (tonnes)	58.2
Average market share for PGMs (%)	9.2

Note: The yearly sales of precious metals are low with respect to the market size and should not disrupt the markets.

time from launch to return in LEO is around 1.3 years. It takes 13 months to build the equipment and the spaceship; total time from investment to the return of the minerals to Earth orbit is thus around 2.4 years. The mass returned to Earth orbit is estimated on the basis of an average concentration of PGMs reported in Table 3.1, the mining season's length; and a throughput of 300. The basic assumptions of the scenario are reported in Table 3.7.

Assuming a Cournot–Nash model with constant marginal costs equal for all firms, the market share of the space mining company is equal to:

$$\chi = \frac{q_1}{q_T} = \frac{1}{(n+1)} \tag{3.1}$$

where

χ = market share of the mining company,
q_1 = yearly quantity sold by the off-world mining company,
q_T = market size of PGMs,
n = number of companies in the oligopoly.

The net present value of this project is given by the following equation:

$$\mathrm{NPV} = \left[\sum \left(p_t q_t - \mathrm{oc}_t q_t\right) e^{-\delta t}\right] - C \qquad (3.2)$$

where:

p_t = price of the metals,
q_t = yearly sales,
oc_t = operational costs, including the transfer of the metals from Earth orbit to surface,
δ = discount rate,
C = investment costs,
t = time from the initial investment.

Price dynamics result from the interaction between supply and demand. We assume a standard demand function:

$$q_d(t) = \alpha p(t)^{\kappa} y_w(t)^{\theta} q_d(t-1)^{\gamma} p(t-1)^{\zeta} \qquad (3.3)$$

We take these quantities to be the firms' strategic variables. Under the supply competition rule, each firm simultaneously submits its supply s_i (p, cp_i, $\mathrm{r}(\mathrm{s_j})$), where cp_i is the cumulative production of firm i, $\mathrm{r}(\mathrm{s_j})$ is the competitors' reaction functions, and p is price. This rule is unrealistic because it is rare that firms have exact knowledge of their competitors' reactions. We change the rule so that the quantity supplied in each period is determined by an adjustment process, in which participants in the oligopoly learn from the historical behaviour of their competitors. The total supply is then a function of prices, cumulative production, and technological factors, though supply adjusts with a lag, and adjustment is incomplete within any period. The intersection of supply and demand determines a market-clearing price. In the simulation, the prices follow a U-shaped curve, as reported in Fig. 3.3.

Unit operating costs are assumed to be equal to labour and administrative costs and the costs of transferring the metals from LEO to Earth. Administrative and labour costs are estimated at 2 % of total gross

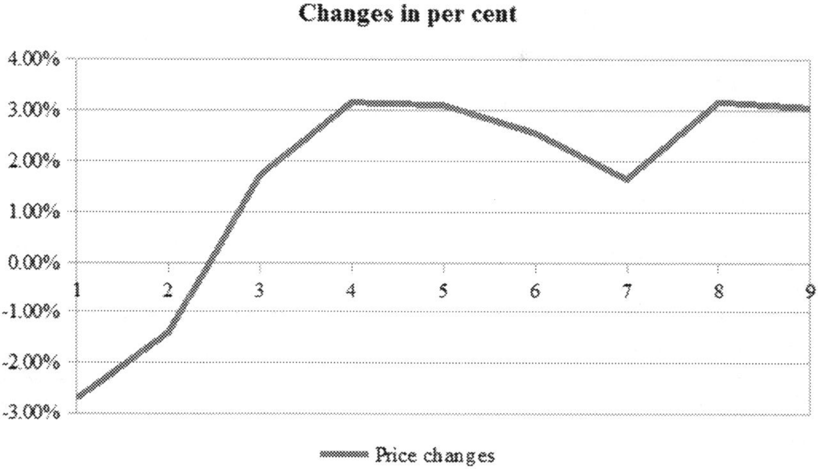

Fig. 3.3 Price dynamics

revenues, while transferring costs are estimated at $5,000 per kilogram. The transfer from LEO to Earth occurs via low-cost space vehicles, such as the Skylon.[20] The results of the simulation are reported in Table 3.8.

These results are based on very simplified assumptions and should be treated with caution, but they offer important indications on project feasibility. First, results depend on the concentrations of PGMs and precious metals. The choice of asteroid is thus decisive for the feasibility of the project. Second, results are sensitive to the discount rate, which reflects two elements: the time value of money (interest rate without risk) and the risk premium. Due to asteroid-mining risk, we choose two discount rates above 15 %. Third, investment costs must be less than the discounted value of sales. The return on investment (ROI) for risky projects acceptable to investors is generally assumed to be over 50 %. We have thus calculated investment costs consistent with ROI over the 50 % minimum (Table 3.9).

[20] The Skylon, a reusable rocket plane, is an initiative of the British company Reaction Engines Ltd. The Skylon reduces the mass ratio by operating in an air-breathing mode in the early stages of the flight before shifting to a pure rocket mode. Skylon's development is estimated to take around nine years and cost around $10 billion. The vehicle has an expected life span of 200 flights. Assuming a production run of 30 vehicles, each vehicle will cost about $774 million. Thus, it should be able to reach a launch cost of about $1,000 per kilogram of payload.

Table 3.8 Simulation results

Equipment mass (in MT)	170
Average price changes for PGMs over the period (%)	2.7
Net value of sales in $ billions	26.6
Period of sales in years	11.0
Discounted net value of sales in $ billions, discount rate 15 %	9.5
Discounted net value of sales in $ billions, discount rate 20 %	7.2

Table 3.9 Rate of return on the investments and investment costs

Investment costs ($ billions)	ROI (discount rate 15 %)	Investment costs ($ billions)	ROI (discount rate 20 %)
6.3	50 %	4.8	50 %
5.3	79 %	4.1	76 %
4.3	121 %	3.5	106 %

In summary, the simulation shows that asteroid mining can be profitable. However, there are two challenges to be faced.

3.3.3.1 Investment Costs

The first challenge concerns investment costs, which at the moment are unknown. Investment costs are driven by the mass to be returned to Earth orbit, mission-velocity requirements, the type of propulsion, equipment costs, and launch costs. Due to its inclination, asteroid 2004 MN4 needs an impulsive transfer and a nuclear engine. A NASA study suggests that the development cost of this technology is between two and a half and three billion dollars. To use an electric-propulsion engine, an asteroid with a lower inclination must be chosen. Research and development of these technologies are well advanced and the next generation of ion engines[21] will offer advanced capabilities with lower development costs compared to nuclear technology.

[21] NASA's Evolutionary Xenon Thruster (NEXT) is significantly improving and extending the current state-of-the-art capabilities of Solar Electric Propulsion Technology. Plasma engines provided propulsion for an attempted landing on an asteroid by the Japanese Hayabusa probe. They were used as well for the trip to the Moon by the European Space Agency's SMART-1 spacecraft.

Launch costs are another important part of investment costs. Whether one uses a nuclear-powered spaceship or an electric one, they have to be brought cold into orbit before starting their engines. Until recently, cartels characterised the launch industry in the United States and Europe, with space agencies dictating the design of the spaceship. Now deregulation has helped in reducing launch costs. Private companies reduce costs through their vertically integrated design and manufacturing, supply chain, and efficiencies in operation. This is already happening in the United States where launch costs are declining, so pressures are building up in Europe to reorganise its launch industry to respond to competition from the United States, India, and China. We are therefore confident that launch costs will decline in a few years time.

J. S. Kargel[22] has analysed the proposed extraction and refining methods for PGMs. These operations are much less massive than on Earth, as they do not need heavy machinery for extraction and transportation of the ore to the refining centre. On asteroids, open cast mining is suitable for regolith. A cap around the mine site can collect the ore, and the cap's rotation, using centrifugal force, channels the mineral to the refining centre. Simple excavators are suitable for hard rock, and a belt conveys the ore to the refining zone. Asteroid minerals need less machining since the metals are already in a free state. The ore is ground and sieved into different sizes, and this material is then subjected to magnetic fields to separate nickel-iron from silicate granules. Electrolytic techniques remove cobalt and nickel from PGMs. Finally, ion exchange techniques, distillation, and solvent extraction will separate the PGMs from gold. Most equipment and costs can be modelled by mining experience here on Earth. The cost of developing specific metal refining technologies in microgravity must also be added. Estimates suggest that equipment costs, including research and development, are around $1 billion, though they can be reduced by building equipment in situ with 3D printing using the asteroid's regolith.

[22] See Kargel J. S. (1994).

3.3.3.2 Legal Issues

The second challenge is legal. Property rights are very important to mining investors. However, the right to mine exclusively an orebody has a higher priority. Note that until asteroids are given property rights, they will be classified as intangible assets, and as such would be subject to impairment testing under International Accounting Standard 36. Uncertainty about property rights and the uses of resources could hinder the funding of these activities by increasing the risk premium. Private companies and financial institutions are not likely to risk their capital and the considerable effort to develop the mineral resources in space, if significant legal or political difficulties might disrupt the supply of these assets.

Under the terms of the "Outer Space Treaty", space, including the Moon and other celestial bodies, cannot be appropriated through claims of sovereignty by any nation. Another clause of the treaty states: "outer space, including the Moon and other celestial bodies, shall be free for exploration and use by all States." If mining is considered to be the use of celestial bodies, experts argue that any material extracted becomes the mining company's property. So it stands to reason that uncertainty over property rights and the use of space resources could hinder financing by increasing the risk premium. The interpretation of property rights is thus important and can only be done by national legal institutions.

The US Commercial Space Launch Competitiveness Act 2015 is intended to clarify these issues since the United States has obligations to regulate its own private space activities under international law. At the end of November 2015, President Obama signed legislation allowing the commercial extraction of minerals and other materials, including water and volatiles, from the asteroids and the Moon. The Act purports to create property rights over resources extracted from asteroids. It states: "any resources obtained in outer space from an asteroid are the property of the entity that obtained such resources, which shall be entitled to all property rights thereto, consistent with applicable provisions of Federal law." The Act defines this entity as "a person or company providing exploration or utilization services that are either organized under US law or subject to US jurisdiction."

This Act received some criticism. Joanne Gabrynowicz, a space lawyer and editor emeritus of the *Journal of Space Law* at the University of Mississippi School of Law, argues that the Act fails to address basic issues, such as who would license and regulate asteroid mining operations, as well as larger issues, such as the legality of mining operations under international law. She argues that current efforts to clarify the legal status of asteroid-mined resources, if approached the wrong way, could guarantee that other nations will challenge unilateral actions and likely cause political competition among nations in space. The reverse is also a concern. Disagreements over space could influence disputes on Earth. In her view, failure to make sure other signatory nations are on the same page could lead to negative geopolitical consequences. If other nations interpret the Act as the United States playing loose with the Outer Space Treaty, they might decide to do so themselves on Earth in unpredictable ways.

Others in the space law community disagree with Gabrynowicz. Brian Weeden is a technical adviser at the Secure World Foundation, a think tank focusing on strategies for peaceful uses of space. He does not expect the sort of conflicts we see over resources on Earth to take place over resources in space, partly because there are vastly larger quantities of such valuables in asteroids than have ever been mined from our planet. He also believes that the Outer Space Treaty is already sufficient legal justification for asteroid mining and that waiting for an international consensus could drag on interminably. He envisions a process where the USA puts the regulatory framework in place in stages to match the staged development of the technology. As it progresses, the USA should engage in a dialogue with other space actors and the international community. If there are issues illuminating particular shortcomings or holes in the international legal framework, they should be addressed then. We agree with Weeden's position.

3.4 The Rationale of Asteroid Mining

Pressure from increasing world population and limited resources is the most cited argument in favour of asteroid mining. For at least three decades, scientists and economists have debated whether humanity is depleting mineral resources.[23] Prevailing opinion is that no immediate concern exists over the current capacity to meet the demand for most metals. Technical progress has extended mineral reserves. The recycling of existing products is adding to the supply of ore. Any related limit to growth is perceived as a long term threat. Consequently, a depleted supply does not offer a rationale for asteroid mining in the short to medium term.

However, there are other important arguments for developing asteroid mining, among which is the criticality of minerals. Critical minerals are referred to as minerals whose supply is constrained while demand is high, and whose market is global in scope. On the supply side, criticality is caused by the confluence of various factors, such as: physical constraints; the influence of non-primary market actors on the market, such as governments; and the fragility of the supply chain. This latter is caused by disruptions arising from operating dislocations. They can be either stochastic (e.g. natural disasters), organizational (e.g. labour unrest and government policies), or institutional (e.g. non-competitive firms' behaviour). Market concentration and other metrics can be first-pass comparators for the criticality level of minerals (Table 3.10).

The first and second metrics consider physical constraints determined by efforts needed to get each kilogram of metal. The costs of extracting

Table 3.10 Metric for criticality

	Ore grade (wt%)	Energy intensity	Prices in US$/kg	Price volatility	% produced in a single country
Cobalt	0.4	0.132	32.2	5.17	53
Nickel	1.5	0.195	19.6	1.97	23
Iron	30–65	0.01	0.14	1.49	24
PGM	0.0003/0.002	240	31,000	3.05	73
Gold	0.005/0.015	200	45,731	1.34	15
Germanium	0.0004/0.001	n.a.	1,950	5.29	68

Sources: Ernst W. G. (2000); Alonso et al. (2007); and USGS (2011)

Notes: Ore grade is usually expressed as a weight percentage of the total rock (wt%). Energy intensity is expressed in gigajoules per kilogram. Non-reliable data are available on energy intensity in germanium extraction

[23] See Kesler S. (2001).

and refining PGMs, gold, and germanium are greater than those of other metals because of low ore concentrations. For example, platinum ore grade varies from 0.5 to 3 grams per metric tonne. This is three orders of magnitude less than the average grade of copper, nickel, tin, zinc, and lead. The energy needed in the refining operations of PGMs and gold[24] is around 1,000 orders of magnitude greater than those for other metals. This explains the high level of PGMs and gold prices. Cobalt, iron, and nickel have higher concentrations and need less energy in extraction and refinement, which explains the lower level of prices.

The largest deposits of PGMs are in countries that are politically unstable or under the influence of non-primary market actors. Hence the fragility of the supply chain to disruptions from operating dislocations is high. In the past, labour unrest and energy shortages in South Africa were the main reasons behind the high price volatility of PGMs. Zimbabwe passed legislation in 2010 requiring large corporations to have 51 % domestic ownership in mining, causing delays in investments to expand existing operations and explore new deposits. This explains the high volatility of the price of PGMs. Nickel, iron, and gold production are more diversified, which makes them less vulnerable to supply chain disruptions. But currently 50 % of cobalt is mined in the Democratic Republic of Congo, a highly politically unstable country, which explains its high price volatility.

The demand for PGMs has increased on average by around 5.5 % per year over the last 60 years. While their demand derives mainly from the autocatalyst and jewellery industries, they are also used in an extensive range of other products and applications. These include glass, electronics, non-transport emission-control equipment, and industrial manufacturing. Collectively they account for 25–30 % of the total demand. The demand for cobalt has increased on average by 5 % per year over the last 60 years and derives primarily from its application as a free metal in the aviation, pharmaceutical, and battery industries. Looking ahead, analysts expect the demand for PGMs to grow at an average annual rate of 6 % over the next ten years, with a major boost for global demand coming

[24] Data on energy intensity to extract germanium are not reliable. Estimates range from 4 to 40 GJ per kg.

from emerging countries, such as China and India. Environmental legislation to reduce auto emissions is expected in these countries over the next ten years. The cobalt market is forecast to grow at an average rate of 6–10 % over the next decade, driven primarily by the demand of the battery industry.

The above analysis shows PGMs are likely to continue experiencing high price volatility. Major factors behind this are an expected high demand, the fragility of the supply chain, variations in energy costs, environmental regulations, and very low concentrations in existing mines. Despite the high energy costs in extraction and refining, gold cannot be considered a critical mineral, as its deposits are diversified in many countries. Nickel and iron are non-critical minerals because of their low energy intensity, their higher concentration in the Earth's crust, and their diversified mineral deposits in many countries. Cobalt is a limiting case though. The low-energy intensity in extraction and refining operations shows that it is not a critical mineral. But the high concentration of deposits in a politically unstable country and the foreseen high demand raise risks of operating dislocations and high price volatility.

During the first years of the twenty-first century, we witnessed major disruptions in the supply chain of PGMs, highlighting the vulnerability of individual companies and/or entire industries. Managers know that they need to protect their supply chains from major disruptions. The most obvious solutions—such as increasing inventories and finding different supply sources—undermine efforts to improve cost efficiency. Recent surveys[25] show managers appreciating the impact of supply-chain disruptions, though they can do little to mitigate its impacts. Managers depend on the innovative ability of producers to replace these metals with new products. Applications of PGMs in industrial sectors—such as the jewellery, electronics, and glass industries—are relatively immune to substitution effects. Because applications in the automotive industry may be affected by solutions based on nanotechnology, many analysts believe the demand for PGMs by that industry will not decrease as a result of substitution, which will instead influence the dynamics of demand by reducing its growth rate.

[25] See Tang C.S. (2006).

The supply of off-world PGMs reduces the fragility of the supply chain, which emanates from political factors. Reduced volatility may thus come from stabler political environments or new supply sources free from political risk. The supply of off-world minerals would be free from such political risk if accompanied by an international legal framework regulating the use of space resources. This would reduce the market volatility risk.[26] Reduced volatility increases the efficiency of the world economy because the automobile and electronics industries are an important part of modern industry.

Will the supply of off-world minerals have welfare benefits? It is known from economic theory that consumers value price stability. Risk averse consumers prefer deterministic consumption streams to risky ones with the same mean. Welfare gains from the supply of off-world minerals depend on the risk involved and the consumers' risk aversion. The prices of catalytic converters mirror platinum, palladium, and rhodium prices. Similarly, the prices of jewellery and electronic products mirror the price of platinum. Reductions in price volatility increases consumers' welfare, although welfare effects are different because the risks involved and consumers' risk aversion vary among these products. The risk is higher for jewellery products than for catalytic converters and electronic products, yet risk aversion is higher for consumers of catalytic converters and electronic products than for jewellery products. The supply of PGMs from space would allow better management of mineral resources on Earth and a longer life for higher concentration deposits, which will increase welfare in producing countries in the medium to long term. In summary, the supply of space minerals makes good economic sense from any economic perspective.

3.5 Asteroid Mining Changes the Geopolitical Framework

Since the end of World War II, there have been more than 150 wars, but only a few were large scale ones between nations. The majority—around 80 %—were civil wars in the developing world. Many scholars have ana-

[26] Since the political instability risk is not easily predictable, the price volatility measure of PGMs (standard deviation) correctly reflects the non-predictable or stochastic price change.

lysed these conflicts to understand why violence occurs and how future conflicts could be prevented. Ongoing debate centres on the geographical distribution of raw materials and how this distribution contributes to conflicts. The distribution of raw materials refers to the geographic occurrence or spatial arrangement of mineral resources on Earth. Metallic minerals are naturally abundant in areas with strong tectonic activity. Fossil fuels are found in sedimentary rocks. Uneven distribution is thus due largely to past and ongoing geological processes.

Such distribution of mineral resources by itself does not necessarily fuel conflicts. The empirical literature has found complex reasons underlying conflicts in the relevant countries. These include access to and control of mineral resources, together with various political agendas. Although weak states are more vulnerable to civil war, there is no single broadly accepted definition of a weak state. A state's strength or weakness can be determined by scaling it according to the following factors: its capacity for coercion; the state of its infrastructure; the condition of its national identity; and the social cohesion within it. Ethnic and/or religious divisions make a country more prone to civil wars, which are often fuelled by various national and international corporations and other regimes with an interest in the outcome of the conflict.

Africa has experienced more civil wars since the 1960s than any other major region around the world. The continent has always been a field of intense economic rivalries among different powerful states, which sought to have access to its natural resources. At the same time, Africa has been a source of strong transnational conflicts over borders and economic, racial, and religious issues. For these reasons, it has experienced extremely violent and bloody civil strife. In several African countries, the abundance of strategic commodities—oil, diamonds, and other strategically important minerals—fuels conflicts. The experiences of countries such as Angola, the Democratic Republic of Congo, Nigeria, Sierra Leone, and Sudan confirm this fact. But experience suggests that no single cause explains these phenomena. Key aspects behind civil wars in Africa are autonomous armed forces, wars within wars, control of the population through fear, external state involvement, political mobilisation on the basis of identity either ethnically or religiously, and the presence of predatory elites.

An example of the complex raft of reasons causing conflicts is the current situation in Sudan, which has been suffering decades of civil conflict as the Arab Muslim tribes in the North are in constant confrontation with the Animistic and Christian African residents of the South. There are significant economic issues as around 80 % of the total oil reserves are located in the South. Until recently, the central government exploited the oil revenues because only the northern part possesses refineries, the critical pipeline, and the terminals necessary to export the oil via the Red Sea. In 1983, a civil war broke out resulting in 2.2 million deaths. In 2005, two regions reached a peace agreement granting autonomy to South Sudan. Darfur is a region in western Sudan with significant gold and uranium reserves and a massive underground water source. It is also an area of fierce ethnic tension. The province of Southern Kordofan, a part of Sudan near the border with South Sudan holding rich oil reserves, is a disputed one and is calling for independence from the North. It seeks union with South Sudan as it has mostly African populations who supported the "People's Liberation Front of South Sudan" during the civil war.

The collusion among local predatory elite, foreign governments, and global companies often further stoke these conflicts. Since 2001, China and Western countries have been engaged in geopolitical competition in Africa. Beijing has taken a series of initiatives aimed at securing strategic sources of raw materials in sub-Saharan Africa and uses its vast international reserves to extend loans in dollars to African countries. This policy helps gain access to Africa's wealth of vast raw material. It displaces the typical game control of the West through the World Bank and the International Monetary Fund. Loans have been made to finance infrastructure and China committed more than $8 billion alone to Nigeria, Angola, and Mozambique in 2006. In comparison, the World Bank extended around $2.5 billion to sub-Saharan countries. Through such moves, China is displacing the Western powers in the game of diplomacy. But aid programmes have brought corruption. The political elites, often conniving with transnational companies, pass along little of economic value to their populations.

The interventions of China and Western powers have gone beyond the standards of international aid as they enable the struggles between various factions in African countries through the arms' trade. Arms went to factions seeking control of strategic minerals to enrich themselves and the armies they use to keep control. Evidence of this trade can be readily found by looking at military spending in sub-Saharan countries, which has increased from around $12 billion in 2001 to more than $18 billion in 2013.[27] Moreover, geopolitical competition for control of strategic minerals contributed to the increased military spending of China and Western powers. They face each other in Africa in an incipient Cold War climate.

In conclusion, Africa's political and economic instabilities are caused by the inability of African countries to form stable governments, a high rate of corruption, the intensification of religious and racial differences, and the collusion of predatory elites with foreign entities. As local governments weaken, extremist Islamic groups intervene. They seek to attract new members for their jihad and to make political as well as economic gains. In some cases, these organisations have developed a high level of collaboration with al-Qaeda offshoots, with the goals of upgrading their profile and diversifying tactical methods. In short, they have exacerbated ethnic and religious conflicts.

Supply of off-world minerals free from political risk could reduce tensions in sub-Saharan Africa, allowing Western countries and emerging ones, such as China and India, to diversify their supply sources. The supply of off-world minerals could thus mitigate terrestrial aggression that seeks the control of mineral resources. These developments may help to reform political systems, including the social, administrative, and legal reforms necessary for their national economic development, as well as the population, and not just the interests of the current predatory elites. These reforms and greater political stability will strengthen the local markets and stimulate the entrepreneurial spirit of these populations, which is needed in Africa to speed up growth. Political stability, less corruption, and higher growth rates help in reducing the attractiveness of terrorism. The above developments may result in a more peaceful international environment and in a gradual reduction in global defence spending.

---

[27] See Stockholm International Peace Research Institute (2014).

3.6 Appendix: Technical Notes on the Net Present Value

This appendix focuses on the determinants of the hypothetical case for net present value:

$$\text{NPV} = \left[\sum \left(p_t \times q_t - \text{oc}_t \times q_t\right) \times e^{-\delta t}\right] - C$$

Following Sonter, the volume of metals brought back from the asteroid to LEO ($\sum q_t$) are derived from the Tsiolkovsky rocket equation, which relates Δv with the effective exhaust velocity and the initial and final masses of a rocket. As the space vehicle's mass is small compared to the mass of the metals, the latter is equal, with a certain approximation, to:

$$\sum q_t = M_{\text{produced}} \times e^{-\Delta v / v_e} \tag{3.4}$$

where M_{produced} is the mass produced on the asteroid, Δv is the change in velocity required for the return by orbital transfer, and v_e is the exhaust velocity.

The mass produced on the asteroid is equal to:

$$M_{\text{produced}} = M_{\text{mpe}} \times f \times t_{\text{stm}} \tag{3.5}$$

where M_{mpe} is the mass of the equipment, f is the throughput factor expressed as the ratio of kilograms per day of the materials treated per kilogram of equipment, and t_{stm} is the time of stay on the asteroid (mining season). Substituting (3.5) into (3.4) gives:

$$\sum q_t = \left(M_{\text{mpe}} \times f \times t_{\text{stm}}\right) \times e^{-\Delta v / v_e} \tag{3.6}$$

This equation shows the mass of the metals returned to LEO and depends on technical factors, the length of the season, the propulsion system adopted, and the change in velocity necessary for the transfer between the two orbits.

It remains to find Δv, the change in velocity needed for the return trip. Δv is a scalar quantity dependent on the desired trajectory and not on the mass of the space vehicle. It is calculated as the sum of the Δv's required for the propulsive manoeuvres during the mission and determines how much propellant is required for a vehicle of a given mass and propulsion system. The Δv_r to insert into the return trajectory is small, due to the low gravity of NEAs, but Δv_{ic} to correct for inclination change must be added.

The asteroid in the hypothetical case is an Aten type. This mission profile assumes a target body with a semi-major axis less than 1, a quasi-circular orbit (an Aten), and a mining season commencing at perihelion and running until aphelion. This mission assumes a Hohmann transfer to rendezvous with the target asteroid at its perihelion and with a near-aphelion departure after half an orbit stay time. The Hohmann transfer orbit is elliptical and used to transfer objects between two circular orbits of different radii in the same plane. An example of the Hohmann transfer is illustrated by Fig. 3.4.

Following Hohmann,[28] the minimum change in velocity for the return trip is equal to the difference between the required velocity to enter the transfer orbit (v_1) and the transfer ellipse velocity (v_2):

$$\Delta v_r = (v_1 - v_2) \tag{3.7}$$

where:

$$v_1 = \sqrt{}\left(\mu / \left(\left(1.5 \times 10_{11}\right) \times Q\right) \times \left(1 - e_{\text{transfer}}\right)\right) \tag{3.8}$$

and

$$v_2 = \sqrt{}\left(\mu / \left(\left(1.5 \times 10_{11}\right) \times Q\right)\right) \times \left(1 - e_{\text{tr}}\right) \tag{3.9}$$

where:

[28] See Thomson (1986).

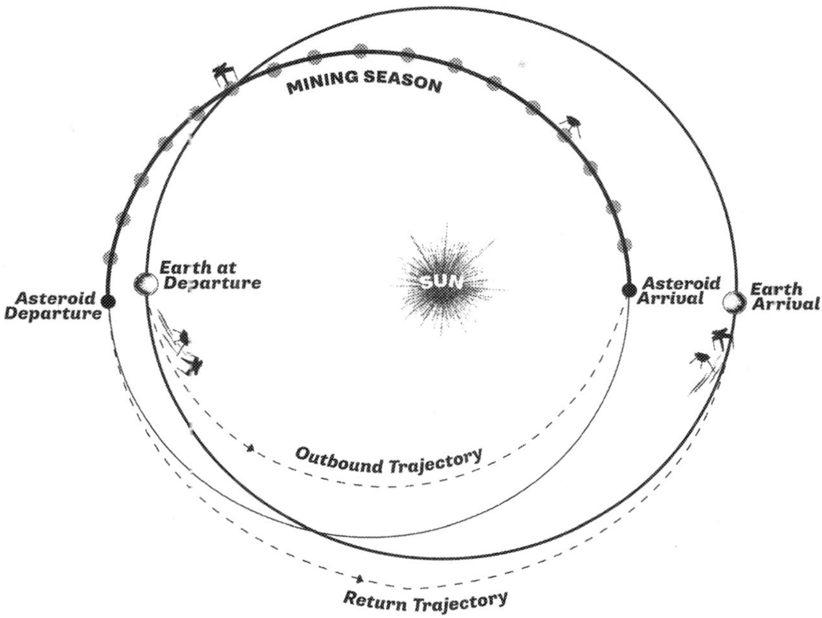

Fig. 3.4 Mission to an Aten asteroid

$\mu = G \times m_s$,
G = Newton's gravitational constant,
m_s = mass of the Sun,
Q = the aphelion of the asteroid in AU,
$e_{\text{transfer}} = (1 - q)/(1 - q)$,
q = the perihelion of the asteroid in AU,
e_{tr} = the eccentricity of the asteroid.

Using the orbital characteristics of asteroid 2004 MN4, v_1 is equal to 37.62 km/s; v_2 is equal to 36.9 km/s. Thus the change in velocity for transfer orbit insertion is 0.73 km/s. However, we have to add the velocity change required for the inclination change, which depends on where the change occurs and the heliocentric velocity at that point. This velocity change can be approximated by:

$$\Delta v_{\text{ic}} = (0.52 \times \text{inclination}) \tag{3.10}$$

which, in the hypothetical case, is equal to 1.73 km/s. The total velocity change for the return to Earth transfer is then equal to:

$$\Delta v_T = \Delta v_r + \Delta v_{\text{ic}} \tag{3.11}$$

or 2.46 km/s.

We assume that the fuel needed for the return trip is extracted from the asteroid. Let us assume the use of a nuclear thermal rocket, which were being actively developed under the NERVA project. Conceptually, this works by pumping hydrogen through the core of a nuclear reactor to heat it to produce exhaust as a reaction fluid. A full-scale operating nuclear rocket was tested and achieved a specific impulse (Isp) of 850 seconds and nearly attained flight readiness before funding was withdrawn. We assume that such a rocket achieves an Isp of 850 seconds. The exhaust velocity is around 8.3 km/s. as the exhaust velocity is equal to the specific impulse times the standard acceleration of gravity. If one assumes that the mass of the spaceship is negligible with respect to the mass of the payload, from the rocket equation, the following holds:

$$M_{\text{start}} = M_{\text{finish}} \times e^{\Delta v / v_e} \tag{3.12}$$

In our hypothetical case, M_{finish} is about 800 tonnes and $e^{\varnothing v/v_e}$ is 1.36. The amount of fuel needed for the return trip is about 320 tonnes. The total time of the mission is equal to the time needed for research and development and the construction of the equipment and the spaceship (t_1), plus the transfer time from LEO to the asteroid and return to LEO. We assume that t_1 is about 1.1 years, and the time from LEO to the asteroid and return to LEO is approximately equal to $3T/2$, where T is the orbital period of the asteroid. In our hypothetical case, this is equal to 1.3 years and the total time of the mission is 2.4 years.

We are finally able to rewrite the equation for the net present value of the project as:

$$\mathrm{NPV} = \left[\sum n_{t=t1-3T/2} \left(\left(p_t - \mathrm{oc}_t \right) \times \left(\left(\left(M_{\mathrm{mpe}} \times f \times t_{\mathrm{stm}} \right) \times e^{-\Delta v_{return}/v_e} \right) / n \right)_t \right) \times e^{-\delta t} \right] - C \qquad (3.13)$$

under the assumption that the investment takes place in year 0, and of constant sales for n years.

Equation (3.13) shows that the total mission time is crucial to the feasibility of the project. The longer the mission time is, the lower the net present value of the project. Δv for rendezvous with the asteroid enters into this equation through C because of the fuel costs for the rendezvous transfer. This equation indicates that an Aten asteroid is most likely the best target for mining. Other types, such as Apollo and Amor, need a longer transfer time and shorter mining seasons. An example of a Hohmann transfer for an Apollo asteroid is given in Fig. 3.5.

These missions imply Hohmann transfers with: a rendezvous near but before aphelion for least changes in velocity; a short aphelion-centred mining season; and a post-aphelion departure for Earth-return, for least

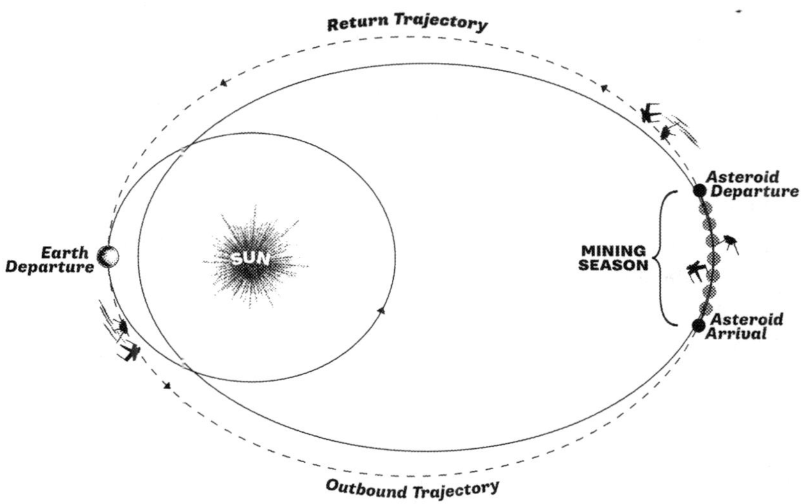

Fig. 3.5 Mission to an Apollo type of asteroid

Δv_r. Such missions have a longer project duration with respect to Aten missions. The use of a Hohmann ellipse transfer and short mining season imply that mission duration approximates to the orbital period of the asteroid. Transfer time is normally longer for Apollo asteroids than Aten ones. Longer project durations affect negatively the net present value. Shorter mining seasons need tougher demand specifications for the equipment for the same mass of metals returned to Earth orbit, which affects the investment costs.

References

Aldrin B. Mission to Mars: my vision of space exploration. Washington DC: National Geographic; 2013. Amazon Books.

Alonso E et al. Material availability and the supply chain: risks, effects, and responses. Environ Sci Technol. 2007;41:6649–56.

Beauford R. Mining the moon for rare earth elements. 2011. http://www.rareearthelements.us/ree_in_space

Buchwald VF. Handbook of iron meteorites: their history, distribution, composition, and structure. Berkeley: University of California Press; 1975.

Cox DW. Islands in space: the challenge of the planetoids. Philadelphia: Chilton Books; 1964.

Craig JR et al. Resources of the earth. Englewood Cliffs: Prentice-Hall; 2001.

Dyson FJ. Disturbing the universe. New York: Basic Books; 1979.

Ernst WG. Earth systems: processes and issues. Cambridge, MA/New York: Cambridge University Press; 2000.

Hoashi M et al. Palladium, platinum and ruthenium in iron meteorites and their taxonomic significance. Chem Geol. 1993;106(1993):207–18.

Kargel JS. Metalliferous asteroids as potential sources of precious metals, J Geophys Res. 1994 Oct;99(25):21129–41. No E10.

Kesler S. Mineral resources, economics, and the environment. New York: Macmillan; 1994.

Kramer G et al. The composition of Kreep: a detailed study of Kreep Basalt 15386. Lunar and Planetary Science XXXIV, 34th Annual Lunar and Planetary Science Conference; 2003 Mar 17–21; League City, Texas; 2003.

Kuck DL. The exploitation of space oases. Princeton Conference on Space Manufacturing, Space Studies Institute; 1995.

Lewis JS. Mining the sky. New York: Helix Books; 1996.
Malvin DJ et al. Chemical classification of iron meteorites. Multi-element studies of 43 irons, resolution of group IIIE from IIIAB, and evaluation of Cu as a taxonomic parameter. Geochim Cosmochim Acta. 1984;48:785–804.
Mankins JC. Space solar power: the first international assessment of space solar power, opportunities, issues and potential pathways forward. Paris: International Academy of Astronautics publications; 2011.
Mankins JC. SPS-ALPHA: the first practical solar power satellite via arbitrarily large phased array. 2011–2012 NASA NIAC Phase 1 Project, Final Report; 2012 Sept.
Mueller RP et al. Regolith as a resource in solar system human and robotic exploration. Paper presented at the 61st International Astronautical Congress. Prague: International Astronautical Federation; 2010.
Muller TG. Thermal infra-red observations of asteroid (99942) Aphophis with Herschel. Astron Astrophys. 2014 Apr.
Pergola P. Small satellite survey mission to the second earth moon. Adv Space Res. 2013;52(9):1622–33.
Sonter MJ. The technical and economic feasibility of mining the near-earth asteroids. Acta Astronaut. 1997;41(4–10):637–47.
Stockholm International Peace Research Institute. Yearbook 2014. Oxford: Oxford University Press; 2014.
United States Geological Sesrvices. Mineral commodities summary, 2011. Washington DC: U.S. Government Printing Office; 2011.
Tang CS. Robust strategies for mitigating supply chain disruptions. Int J Logist Res Appl. 2006;1:33–45.
Thomson WT. Introduction to space dynamics. New York: Dover; 1986.

4

The Great Leap Out of the Cradle: The Human Exploration of Our Solar System

Moving human space exploration beyond the Moon is the next challenge. Mars, the most Earth-like planet in the solar system, is the next target. Table 4.1 summarises the main physical characteristics of the Earth and Mars.

Mars has a thin atmosphere[1] and low average surface temperatures (between −140 and 20 °C). Carbon dioxide is the main component of the atmosphere at 95.3 %; nitrogen and oxygen are around 3 %. Mars offers limited protection from cosmic rays due to the absence of a planetary magnetic field. Water exists as ice, with a small quantity present in the atmosphere as vapour. Water ice is visible at the surface at the north pole and abundant frozen water may be present beneath the permanent carbon dioxide ice cap at the south pole. It may also be present in the shallow subsurface at more temperate latitudes. Martian soil is slightly alkaline and contains chemical elements such as magnesium, sodium, potassium, and chlorine. These nutrients are found on Earth and are

[1] The pressure at the ground is less than one-hundredth of the atmospheric pressure on Earth and has much variability with altitude and latitude. It varies from around 0.3 millibars on Olympus Mons to over 11.6 millibars in the depths of Hellas Planitia. It has a mean surface level pressure of 6.36 millibars, ranging from 4.0 to 8.7 millibars depending on the season.

G. Bignami, A. Sommariva, *The Future of Human Space Exploration*,
DOI 10.1057/978-1-137-52658-8_4

Table 4.1 Main characteristics of Earth and Mars

	Earth	Mars
Mass (10^{24} kg)	5.9736	0.64185
Volume (10^{10} km^3)	108.321	16.318
Equatorial radius (km)	6378.1	3396.2
Polar radius (km)	6356.8	3376.2
Ellipticity	0.00335	0.00648
Mean density (kg/m^3)	5515	3933
Surface gravity (m/s^2)	9.81	3.71
Surface acceleration (m/s^2)	9.78	3.69
Escape velocity (km/s)	11.19	5.03
Solar irradiance (W/m^2)	1367.6	589.2
Orbital semi-major axis (10^6 km)	149.60	227.92
Perihelion (10^6 km)	147.09	206.62
Aphelion (10^6 km)	152.10	249.23
Mean orbital velocity (km/s)	29.78	24.13
Orbit inclination (deg)	0.000	1.850
Orbit eccentricity	0.0167	0.0935
Length of day (hours)	24.0000	24.6597
Axis rotation inclination (deg)	23.45	25.19

necessary for plant growth. So, while Mars is not a suitable holiday place, it is not impossible for humans to live there.

Since the Apollo programme ended, human exploration of Mars has been a goal of space agencies and other organizations over the intervening decades. In fact, these organisations have been putting forward mission proposals[2] since the 1950s, but none of them has gone beyond the study stage. However, this long period has seen some benefits since it allowed: ideas on the architecture of such a mission to mature; misconceptions to be overcome; and research on necessary technical innovations to take place. But a fundamental flaw may have been that, while technical details and the architecture of the mission were extensively considered, the political, economic, and financial motivations were not sufficiently explored.

[2] Mission proposals include among others: (i) Project Empire, a study commissioned by the Marshal Space Flight Center and awarded to General Dynamics and Lockheed (1959); (ii) von Braun's Mars Project (1969); (iii) The Martian Expeditionary Complex of the Soviet Union (1969); (iv) the NASA Space Exploration Initiative (1989); (v) Mars Direct by R. Zubrin and D. A. Baker (early 1990s); (vi) the NASA Design Reference Mission (early 1990s and 2000); (vii) the Mars Piloted Orbital Mission of the Russian Space Agency (2000/2005); (viii) the Aurora Programme of ESA (2001); the ESA–Russian Plan (2002); and the Vision for Space Exploration of the United States (2004).

This chapter looks at Mars exploration and settlement by developing a specific scenario, which assumes that near-Earth asteroid exploration is accomplished in the early 2020s and, soon after, asteroid mining is carried out. It assumes that the core of the space station for the storage of minerals and fuels, extracted from the asteroids, is completed in the 2020s. We first analyse the technical feasibility of the human exploration of Mars; second, we analyse the costs and how to finance this venture; third, we analyse the technical, economical, and institutional feasibility of establishing a permanent human base on Mars. Finally, we look at the possible evolution of the exploration of the solar system beyond Mars.

4.1 The Feasibility of Mars Exploration

Defining the architecture of the mission is a difficult problem, because of the need to define possible options and to assess criteria for the scenario. Key possible options related to a Mars exploratory mission by humans includes: (i) mission goals; (ii) the landing site; (iii) the Mars-outpost base camp and logistics; (iv) interplanetary propulsion; (v) the configuration of spaceships; (vi) the interplanetary trajectories of systems; (vii) entry-descent and landing strategy; (viii) space-station strategy; and (ix) telecommunications. The assessment criteria for the scenario are technological readiness, cost efficiencies, safety of the astronauts, and organisational efficiency. Some of these criteria may be contradictory, and when this occurs mission goals and astronauts' safety are the ultimate determining criteria in our choices.

4.1.1 Mission Goals

Successful proposals should aim at optimising the chances of the project's approval by defining goals subsuming both forceful political and economic motivations. Mars exploration proposals of the last decade highlighted scientific goals, including the search for life outside Earth and the positive economic effects. These goals make a powerful tool in the hands of governments to push forward innovation capabilities and economic growth. But history teaches us that they were not enough, and human Mars exploration has remained stalled at the proposal stage.

The main long-term goal should be to build a permanent human base on Mars. Commitment to this goal is more likely to get support by the public, policy makers, and private companies. We believe that colonisation of Mars will rekindle humanity's propensity to explore. The desire to explore and migrate has characterised humanity since the dawn of its history. Since *Homo sapiens* stepped out of Africa and expanded into Eurasia and Oceania, 30,000 years have passed. Humans populated the Americas, from Alaska to Chile, over some 2,000 years, from 16,000 to 14,000 years ago. During the early history of our species, most likely resulting from a variable climate during the last part of the Pleistocene and an inadequate food supply, rather than voluntary conscious motivation, people moved at a slow pace following animal herds, which also migrated on account of changes in the climate. Given the vast distances and primitive means of travelling and communication, several human communities emerged, isolated and adapted to different environmental conditions. Different cultures arose because of their isolation. Today, anthropological and genetic studies show that humans differ more in their languages and cultures than in their genes.

Recent research has identified a variant of the DRD4 gene that controls dopamine, which facilitates transmission of signals between neurons and plays a major role in the brain network responsible for attention and reward-driven learning. This variant, the DRD4-7R allele, actually blunts dopamine signalling, which enhances individuals' reactivity to their environment. It also drives the search for the new and for risk taking. This variant evolved in the distant past, enhancing the survival prospects for those who possessed it. Guy Gugliotta [3] has suggested that resource depletion or changing environments might have selected for individuals with this variant. Such adaptation most likely played a role in the out-of-Africa event. Later on, local cultural environments influenced allele frequencies associated with such behaviour. Preliminary results of this research show this allele present in 20 % of humanity, and this number may eventually increase due to ongoing research.

[3] See G. Gugliotta (2008).

Inclinations for discovery and exploration are ingrained in the human mind or at least in one mind out of five. But scientists propose that human behaviour is determined also by the environment, that is by the available means and opportunities. The history of exploratory ocean voyages between the fifteenth and nineteenth centuries teaches us that scientific discoveries and technological advances (means) were instrumental for these voyages. Collaboration between governments and individuals, each pursuing their own strategic interests (opportunities), characterised these voyages. Competition for power among nations was also a powerful motivation. Often governmental interests joined those of individuals in the search for knowledge, adventure, and/or escape from political instability and of companies looking for new trade routes, raw materials, and outlets for their products.

Voyages of exploration preceded migrations by creating the knowledge and infrastructure that then allowed future settlement. From the fifteenth century onwards, humanity has undergone an integration which today we call a global society. Migrations started in the eighteenth century (including involuntary migration in the form of the slave trade) and by the nineteenth the rise of the European industrial empires contributed further. Several countries favoured their own ethnicities over outsiders, but others appeared to be more welcoming. Transnational labour migration reached a peak of 3 million migrants per year in the early twentieth century. Italy, Norway, Ireland, and China's Guangdong region were places with high emigration rates. And the pace of migration has sped up today. People leave their home countries looking for a better life and to escape from political turmoil.

Humans have now populated every habitable corner of Earth, which has become a single habitat. However, the planet may reach a tipping point. According to recent research, projections by the Club of Rome match current data that estimate a high likelihood of environmental collapse by 2050 if present trends remain unchanged. Without game-changing events or breakthrough technologies, humanity will be forced to confront its "limits to growth." Even though we are steadily consuming the Earth's finite endowments, new technologies offer hope by creating alternatives and increasing efficiencies. Nevertheless data show that annual global per-capita consumption patterns continue to increase, which is an irreversible path that could lead to societal collapse. However, utilisation of the space arena might avoid the realisation of the Club of Rome's predictions.

Mars colonisation would expand and enlarge the space economy beyond the Moon and near-Earth asteroids, increasing trade opportunities and contributing important innovations to the Earth economy. The base would offer goods and services to spaceships travelling between Earth and Mars and to those venturing into space beyond. Current indications are that Mars and its two moons are rich in hydrogen, the fuel required by nuclear-powered spaceships. While fuel costs are great if sent from Earth, Martian hydrogen would offer a competitive supply. Another resource that abounds on Mars is deuterium, a heavy isotope of hydrogen. It occurs at a rate of 160 out of every million hydrogen atoms on Earth, but at a rate of 833 on Mars.[4] Deuterium is the key fuel for spaceships with nuclear-fusion propulsion. In addition, it could be an exportable commodity for sale and use on Earth. Contrary to common thinking, deuterium is an expensive raw material on Earth with a value of around $10,000 per kilogram which is 50 times more precious than silver. Once nuclear-fusion reactors become commercial on Earth, its price will increase and make deuterium a marketable product.

Agricultural products are valuable commodities for exchanging with spaceships travelling between Earth and Mars, and beyond. When space mining has been extended to the asteroid belt, miners will need a steady supply of food, which must be sent either from Earth or from Mars. The latter has a decisive competitive advantage.

The maintenance of spaceships is another marketable resource. Mars has the minerals necessary to produce the components of spaceships. Because of its complex geological history, critical and precious metals may exist on Mars. At the moment, the presence of these metals is hypothetical, but it could become real once settlers have had the chance to find them. In addition, Mars's moons are also a likely source of precious metals. Companies engaged in asteroid mining would extend their

[4] A recent study compared Mars's atmospheric concentrations of water to heavy water. It discovered that the concentration of deuterium over Mars's polar ice caps is now much higher than in Earth's oceans. See Villanueva G. L. et al. (2015).

operations to Mars and/or to its moons, and the precious metals they extract would be valuable goods for exchanging with Earth.

Another product marketable on Earth is innovation. It is known from history that a frontier society stimulates innovation for its own survival. Labour shortages prevailing in the colonial era of nineteenth-century North America spurred "Yankee ingenuity," which, conjoined with a richly technological culture, would drive the ingenuity of Martian *Homo sapiens* to produce waves of inventions that would predominately be in energy production, automation, robotics, biotechnology, and medicine. These inventions, patented on Earth, could finance the base on Mars and revolutionise and raise Earth's living standards just as nineteenth-century American inventiveness changed Europe and the rest of the world.

Establishing a permanent human base on Mars will need careful planning, including an initial human exploratory mission. The main goal of this mission would be to explore the feasibility of human settlement on Mars and gather and analyse data for the base's location. Scientific goals are not excluded, but they are secondary to the main goal. The mission's goals help in defining its architecture's main parameters. At least four robotic spaceships must precede the human mission and follow a low-energy transfer trajectory. These spaceships will contain fuel for the return trip, a descent/ascent vehicle, pre-assembled materials necessary for the astronauts' survival, and the tools for exploration. Once this has been done, the human mission composed of at least six crew members will depart and, when in Mars orbit, a subset of astronauts will descend to the surface. The others will stay on board the spaceship to carry out scientific experiments and survey Mars and its moons from orbit. They will also manage the communications between Mars and Earth. The stay time on Mars will depend on the mission goals, so it may vary from 40 to 500 days.

4.1.2 The Landing Site

There are four criteria for choosing a landing site: the presence of water; the availability of building materials; safety issues; and, most importantly, the mission's goals. Preliminary scouting for potential sites of a permanent human base will weigh heavily on the choice of landing site. A permanent human base should be either underground or in existing caves since conditions on the surface are too hostile for long-term human residence. We now review the pros and cons of five potential sites.

Tharsis is a vast volcanic plateau centred near the equator in Mars's western hemisphere. It is home to the largest volcanoes in the solar system, which include three enormous shield volcanoes: Arsia Mons, Pavonis Mons, and Ascraeus Mons. Pavonis Mons hosts several lava caves. Images from orbiting vehicles have identified skylights or breaks in the top of buried lava tubes, which would offer shelter from radiation[5] and provide a constant ambient temperature. This site presents few drawbacks, but access to the lava caves would be challenging, as they are at a high elevation. Moreover, unlike the northern plains, Mars's lava tubes might not have water.

Arabia Terra is a northern highland with dense impact craters dating back billions of years and canyons. Many canyons terminate at the northern lowlands, bordering Arabia Terra to the north. Craters offer protection from winds, which can be very strong and full of dust. Water, most of it bound within minerals, is present in this region. In terms of energy and technology, water will be complicated and costly to extract, but it might be present inside a crater that was made by the impact of a comet on the Martian soil.[6] Images show an iced lake a few kilometres in size and with an estimated depth of 100 metres. The major drawback of this landing site is the lack of protection against radiation.

Hellas basin is located in the southern hemisphere and has many glaciers which are fossil remnants with iced water at shallow depths. As in Arabia Terra, no natural shielding exists against radiation. A fourth site is composed of the canyons of *Valles Marineris*, where spring water deposits may exist as it periodically breaks through ice plugs. Daytime temperatures can be around zero degrees Celsius. While surface water

[5] Mars has a weak magnetic field, present over large areas of the planet, but that does not extend over the entire surface. Hence, Mars is not protected as Earth is from cosmic radiation. We do not have good information on ultraviolet radiation that reaches the Martian surface. Consequently, a more detailed understanding of the radiation environment is needed. Present robotic exploration should be collecting the information necessary to assess the effects of radiation on astronauts and future colonists.

[6] ESA's Mars Express mission identified this crater.

may exist only a couple of times a year, the liquid aquifers may be accessible, though reaching them up slopes hundreds of metres above the landing site would not be an easy job.

The last potential site is at the *outer edge of the North Polar Cap*. Here water exists close by, while this flat region is safe for landing. Mars's North Pole is 6 kilometres lower in elevation than the South Pole which offers more time to slow down landing craft and correct their positions. Since Mars's solar orbit is greater than Earth's, its northern summer has nearly six months of continuous daylight. This allows direct-to-Earth communications most of the time, in contrast to non-polar sites, which have to rely on links through orbiters.

This comparative analysis points to a landing site at a safe distance from the geologically interesting regions of Valles Marineris, the Hellas basin, and the Tharsis region (Fig. 4.1).

4.1.3 The Mars Outpost Base Camp and Logistics

The Mars base camp is designed to test the feasibility of long-term human stays on Mars and would consist of a central habitat, a life science module, and the power systems.

4.1.3.1 The Central Habitat and Life-Science Module

The habitat and the life-science module must have solid wind and radiation protection. They also need to be pressurised and heated. Wind protection is necessary because winds on Mars can be very strong and full of dust which is abrasive and toxic. The dust contain corrosive oxides and hexavalent chromium ions. Cosmic radiation protection is also needed because Mars has a weak magnetic field and a thin atmosphere which offers poor protection against solar radiation at the surface. Radiation doses in sieverts per year are estimated to range between 0.2 and 0.3. At low altitudes, the atmosphere does provide better protection.

Several proposals for the habitat's architecture exist and are categorised into two classes. Class I are pre-fabricated structures deployed on

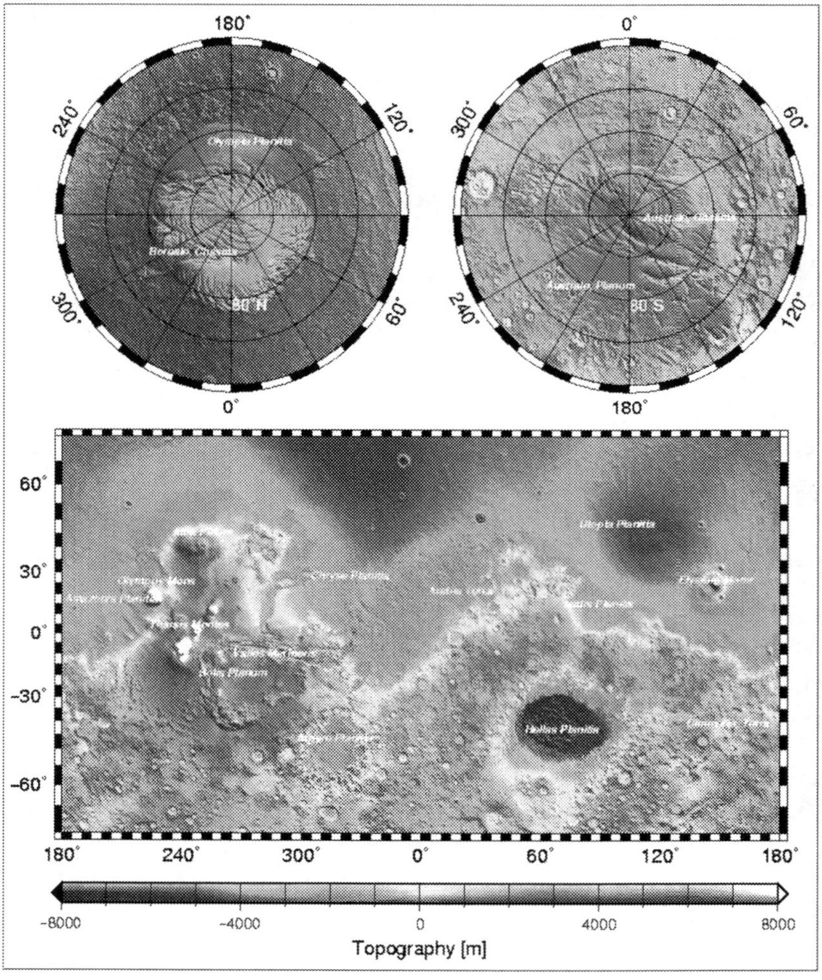

Fig. 4.1 Topography of Mars (*Source*: NASA (http://mola.gsfc.nasa.gov/images.html). *Notes*: North is at the top. Notable features include the Tharsis volcanoes in the west (including Olympus Mons), Valles Marineris to the east of Tharsis, and the Hellas basin in the southern hemisphere)

the surface. One of these is the Transit Habitat or Transhab, which is a unique hybrid infrastructure that combines a hard central core with an inflatable exterior shell. The central core is the hard part comprised of longerons, repositionable launch shelves, bulkheads, radiation shield water tanks, utility chases, and integrated ductwork. The inflatable shell is composed of four layers: the internal scuff barrier and pressure bladder, the structural restraint layer, the micrometeorites shield, and the external thermal-protection blanket. An integrated pressurised tunnel is located at one end to enable access to the surface. A tunnel, not pressurised, is located at the opposite end, which houses the TransHab inflation system. Because TransHab is prefabricated, packaged, and deployed, it requires the crew to set it up.

Class II includes structures manufactured using local resources. One of these is the Adobe[7] infrastructure made with Martian regolith, which is a free resource available in unlimited quantities on the surface. The Adobe exterior serves as a radiation shield and a thermal insulation layer and provides complete protection against dust storms and micrometeoritic impacts. It houses an inflatable membrane that functions as a self-contained pressure vessel for the crew to inhabit. Its construction requires on the surface a regolith drill/pump, a brick baker, plastic superadobe bags to hold soil, an inflatable half-cylinder membrane, and a plexiglass sheet supply.

The brick baker is a simple device. It pours regolith into mild compression moulds and a small furnace bakes it to produce bricks. The habitat wall is built using coil and bags filled with regolith. Each layer is placed on top of the lower layers and is hardened by the compression of the layers above. Specific spaces are built into the wall for power and gas lines, plexiglass windows, and airlock junctions. The roof is built using the "leaning arch" technique. After the exterior is built, the crew lays the base floor matting before bringing in the inflatable to match up its airlocks with the wall openings. The membrane is inflated with a mixed oxygen gas.[8] Buffer gases are regulated at 9 psi to supply pressure. Internal

---

[7] Adobe is the Spanish word for mud brick, made from natural building materials such as sand, clay, water, and fibrous or organic material.

[8] While it is possible for humans to breathe pure oxygen, such an atmosphere is dangerous as demonstrated by the fire in the Apollo 1 capsule. So the Mars habitat may need additional gases. One possibility is to take nitrogen and argon from Mars's atmosphere. Separating nitrogen from argon is difficult. It is suggested that the Mars habitat uses 40 % argon, 40 % nitrogen, and 20 % oxygen. Thanks to modern technology, Mars's atmosphere could be compressed and adjusted to form a breathable mixture for humans. But the effects of breathing gaseous argon are unknown. Ongoing experiments are being conducted to monitor the health and activities of different animals while breathing a gas mixture that includes argon.

supplies, furniture, and equipment are brought in after inflation. Two inflatable tunnel airlocks are attached to the habitat, one joining the lander and the other joining the life-science module.

Class I and II habitats offer adequate solutions to protect astronauts against wind, radiation, and low temperatures. The Abode infrastructure serves as a shell to house the pressurised membrane. In contrast, Transhub needs hard structural support to mould and contain the 9 psi environment. The Adobe infrastructure reduces the overall mass that needs to be transported to Mars and the mission's costs and offers useful information and experience for a permanent human base. Attaining cost efficiency and long-term goals rule in favour of Class II solutions.

4.1.3.2 Power Systems

Robust power systems are essential for the survival of the crew and include nuclear or solar power for the base camp, and electrical power for one exploration vehicle and emergency rover. The base camp needs power to operate the life support system, for thermal management, for in situ resource facilities, and for scientific equipment and other devices. The power system should be capable of producing a power level of 200 kilowatts, which is based on four astronauts living in the base and the in-situ resources required to build the habitat and the life science module and water extraction.

Two options are being investigated by space agencies and research laboratories. A first option includes a nuclear reactor coupled with a Stirling engine to generate electricity. A liquid metal[9] transfers the heat from the reactor to the Stirling engine, which uses gas pressure to convert heat into electricity. The nuclear reactor should be small, compact, and safe. One candidate is an SP-100 reactor.[10] The reactor has integral shielding, is self-contained, and includes its own auxiliary power for start-up. The

---

[9] The liquid metal is a sodium potassium mixture, which is used to transfer heat from the reactor to a generator.

[10] SP-100 (**S**pace Reactor Prototype) is a US research programme for nuclear-fission reactors in space. A significant part of the SP-100 Project is aimed at identifying plausible missions and determining their requirements. Preliminary results indicate significant benefits of nuclear-reactor power systems.

reactor, primary coolant, and power conversion components are housed in a vacuum vessel for protection from the carbon dioxide environment. The vacuum is maintained by an active-ion pumping system, which has no moving parts and a long record of reliable operation. Assuming a specific mass value of 150 kilograms per kilowatt hour, the reactor mass is 30 tonnes.

Solar power is a second option. Operating photovoltaic arrays on the surface of Mars is a significantly different problem than operating them on the Earth's surface. Mars is much further from the Sun than Earth and the solar intensity is much lower than on Earth. The suspended atmospheric dust consists of a long-term component constantly aloft in the atmosphere of Mars, which also has dust storms that temporarily increase the amount of dust. This significantly different environment compels the use of a very large surface array of solar cells. The need to have energy available during the whole day-night cycle as well as during dust storms requires efficient energy-storage devices. Dust accumulated on the arrays after a dust storm must be removed, either by robots or by the astronauts themselves. In summary, nuclear power, unlike solar power, provides the constant energy source necessary for the support of human life. In addition, a large solar array and the energy-storage devices would add yet more mass to be delivered from Earth than in the first option. Safety and costs considerations rule in favour of nuclear power.

Rover-enabled exploration focuses on expanding our knowledge of the Martian environment while searching for a potential location for the permanent base camp. Exploration is performed using a long-range traverse vehicle, which is capable of traversing the Martian landscape while providing a comfortable living and working environment. The vehicle is supplied by cargo landers that will place supply caches at several locations along the route. Each cache provides human necessities along with scientific and monitoring equipment. A rover vehicle with a 30 kW user net power is being considered. The operations envisaged as enabled by this power include drilling, sampling and sample analysis, on-board computer and computer instrumentation, vehicle thermal management, and astronaut life support. Besides the 30 kW net user power, electric power is needed to propel the rover across the terrain.

The only system that can offer this level of power at a reasonable mass penalty is a nuclear-reactor heat source with static or dynamic conversion. Current research efforts are focused on the NASA SP-100 reactor. To integrate a nuclear reactor into the long-range traverse vehicle one needs a detailed description of the rover's requirements. A pressurised rover may have the size of a city bus with a mass of around 8 tonnes. The vehicle performance requirements are a velocity of 10 km/hr; a range of 100 kilometres; a slope climbing ability of 30 degrees for 50 kilometres; a single mission duration of five days; and a crew of two. A variety of technical issues need to be resolved, including minimisation of the shield mass for protection of the crew from nuclear radiation, optimisation of the shield configuration, and integration of the heat emitting radiator with the design of the rover. The power system's specific mass of 100 kgs per kwh yields a total vehicular mass of 10 metric tonnes, including 3 tonnes for the reactor based on SP-100 technology. The remaining 7 tonnes of mass is devoted to the radiation shield. The vehicle will have an auxiliary power system (a radioisotope power system[11]) as a back up in case of failure of the primary unit.

One further power source, although relatively minor, is needed. Astronauts need spacesuits to work on Mars; the ideal spacesuit is an exoskeleton, power-assisted, pressurised one that is well-insulated but lightweight and flexible enough to accommodate the astronaut's movements and different from the bulky, unwieldy ones used by the Apollo astronauts. Space agencies are continuing to work on this new kind of spacesuit. An important problem waiting solution is the exoskeleton's energy source. The most promising power solution centres on new electricity accumulators based on nano-capacitors.[12] Nano-capacitors are formed by clusters of platinum atoms surrounded by clouds of carbon atoms. Among the current studies is an interesting experiment on electric-charge accumulation being conducted in Italy at the University of Milan.

[11] Radioisotope power systems provide electricity and heat, enabling spaceships to undertake missions to environments beyond the capabilities of solar power, chemical batteries, and fuel cells. These technologies are capable of producing electricity and heat for decades under the harsh conditions of deep space without refuelling.

[12] Nano-capacitors store electrical charge on two metal-electrode surfaces separated by insulating material. Their storage capacity is directly proportional to the electrodes' surface area. Electrodes work in the same way as in conventional capacitors, but, instead of being flat, they are tubular and located deep inside the nanopores.

4.1.4 Interplanetary Propulsion Systems

Rockets are momentum machines. They spew gas out of nozzles at high velocity causing the nozzles and the rocket attached to them to move in the opposite direction; they are governed by Tsiolkovsky's rocket equation:

$$\Delta v = v_e \ln\left(m_0 / m_1\right)$$

where v_e is exhaust velocity, m_0 is mass including the propellant, and m_1 is mass without the propellant. The equation has three variables, which can be conceived as energies: (i) energy expenditure against gravity (Δv or change in rocket velocity); (ii) energy available in the rocket propellant (exhaust velocity or specific impulse); and (iii) the propellant mass fraction (the quantity of propellant compared to the rocket mass). Given any two of these variables, the third one becomes fixed.

Next, we need to choose the type of propulsion, which specifies the available energy. Mars exploratory proposals have considered two propulsion systems: chemical and nuclear. The Appendix discusses the details of these propulsion systems. Chemical rockets produce thrust by expelling the exhaust at high speed, which is usually a gas created by the high-pressure combustion of the propellants, hydrogen and oxygen, and which passes through nozzles that use the gas's heat energy to accelerate the propellant. The exhaust velocity generated by a chemical engine is limited because the fuel provides both heat and thrust. In contrast, a nuclear rocket maintains a separation between the heat source and the fuel, which allows for greater exhaust velocity with less propellant usage. Table 4.2 shows the efficiency of different engines in terms of fuel as needed for a round trip to Mars.

Estimated velocity changes include Δvs to make a 7 degree inclination change to the Earth orbit.[13] Table 4.2 shows that nuclear propulsion is

[13] We assumed that the mission launch plane is inclined by 28.5 degree to the Earth's equator, but 23.5 degree to the ecliptic plane. The plane change is thus 5 degree. Allowing a window within which to make the burn, the plane change is closer to 7 degree.

Table 4.2 Efficiency of chemical and nuclear engines

	Chemical engine	NERVA Deriv (H2)	Rubbia engine
Minimum Δv (m/s)	7957	7957	7957
Exhaust velocity (m/s)	4462	8085	30,000
Mass of the spaceship without fuel (tonnes)	60	60	60
Propellant mass fraction	4.95	1.68	0.30

Notes: The spaceship's mass without fuel are arbitrarily fixed at 60 tonnes. LEO and LMO altitudes are arbitrarily selected at 400 and 200 km respectively. Velocity changes at departure are computed for the average orbital locations of Earth and Mars and for a short trip, which means that Earth is at aphelion at departure, and Mars is at perihelion at arrival

more efficient than chemical propulsion. Nuclear engines have another advantage over chemical engines since they can achieve higher velocities and reduce travel time. This limits exposure risks to cosmic radiation for astronauts and the spaceship's sensitive components. The drawback is that the spaceship's insertion into Mars orbit requires complicated manoeuvres[14] (deceleration), which requires more fuel mass than the ones shown in Table 4.2. But this choice becomes necessary if radiation during the time spent in space proves to be a show-stopper. Another advantage of nuclear propulsion is that the spaceship, on its return from Mars, will not re-enter Earth's atmosphere from a hyperbolic trajectory. Instead it will stay in LEO or in higher orbits where it can be refurbished and reconditioned for the next Mars mission. Multiple journeys can thus be performed in a more cost-effective way. This is particularly important if the main long-term goal is to build a permanent base on Mars.

Nuclear propulsion is not a fully mature technology and has a lower technology readiness level (TRL) than chemical propulsion. Although the TRL of nuclear engines varies, nuclear thermal propulsion has not such a low TRL and tests performed in the 1970s on a prototype version contributed to the development of safe, reliable thrusters in a reasonable time and with reasonable costs. Nuclear electric propulsion TRL can be

[14] The spaceship needs a specific velocity to orbit Mars. This velocity is less than the one needed to continue to orbit the Sun in the transfer orbit. Hence the spaceship has to decelerate to be captured by Mars's gravity.

split into two parts: the design of the power plant and the design of the thrusters. In the former case, the TRL is advanced. In the latter case, small electric thrusters represent a mature technology. Research aimed at scaling them up and improving their performance is under way.

Table 4.3 shows an estimate of costs for nuclear thermal propulsion, which is based on reviewing and revising the designs of the NERVA programme. The total costs to rebuild a NERVA engine is estimated at $3 billion. The Rubbia engine and other advanced nuclear engines have the lowest TRL, so extensive research and development programmes are needed. R & D cost estimates are not available, but they are likely to be multiples of the R & D costs for traditional nuclear propulsion.

In summary, nuclear propulsion is more efficient than chemical propulsion and it offers the choice of faster velocities and lower exposure risks to cosmic radiation. Chemical propulsion has lower development costs due to its high TRL, but, if many missions are planned, the nuclear propulsion development costs can be considered as an investment that will be repaid by lower costs of the next missions. Based on assessment criteria and mission goals, nuclear is the preferred propulsion.

4.1.5 Interplanetary Trajectories

When travelling between planets, it is necessary to minimise the propellant mass, which depends on the chosen route. To launch spaceships from Earth to Mars using the least propellant, consider that they are already in solar orbit when sited on launch pads. This existing solar orbit must be adjusted to take the spaceships to Mars. The desired orbit's perihelion (closest distance to the Sun) will be at the distance of Earth's orbit, and the desired aphelion (farthest distance from the Sun) will be at the

Table 4.3 NERVA engine research and development costs

	Costs ($ billions)
Design and construction	1.2
Technology	0.4
Capital	0.5
Operating	0.2
Power station test	0.7
Total	3.0

Table 4.4 Parameters of the two missions

	Short stay (months)	Long stay (months)
Outbound	7.5	7.5
Stay	1.2	15.3
Inbound	9.7	7.7
Total time in space	17.2	15.4
Total mission time	18.4	30.7
Propellant (metric tonnes)	46	39

Notes: The time spent in space and on the planet depends on the launch window. The numbers shown may vary according to the launch window. The estimates of the propellant needed are based on the use of the Rubbia engine. The spaceship mass is 150 tonnes. These estimates assume the minimum Δv in the two missions

distance of Mars's orbit. This is called a Hohmann transfer orbit. The solar orbit's portion taking the spacecraft from Earth to Mars is called its trajectory. Two possible interplanetary trajectories exist: conjunction and opposition trajectories. Several papers[15] have analysed the trade-off between the two trajectories. Important assessment criteria in choosing between them are: efficiency in terms of the propellant's consumption; the transit time; astronauts' safety; and the stay time on Mars. Table 4.4 displays these parameters.

The conjunction-class trajectory is the most efficient, but staytime on Mars is around 500 days. A shorter stay on the planet (40 days) is possible but at the expense of more propellant. Departure from Earth should be performed earlier to rendezvous with Mars, a few weeks or months before the two planets' conjunction. Time in space is longer following opposition rather than conjunction trajectories. Time on the Martian surface is shorter following opposition trajectories. It is doubtful that any mission's goals could be accomplished in such a short period. Dangers to the crew result mostly from time spent in space owing to more radiation exposure and the longer weightlessness periods. In summary, the main mission goals, efficiency, and astronauts' safety rule in favour of conjunction-class trajectories.

[15] See Zubrin R. et al. (1990); George L.E. et al. (1998); and Landau D.F. et al. (2006).

4.1.6 Spaceship Configuration

There are different possible configurations for manned interplanetary spaceships, cargo and descent/ascent vehicles, and landing vehicles

4.1.6.1 Manned Interplanetary Spaceships

Our choice is to use the same manned interplanetary vehicle for the inbound and outbound trips. Although this configuration can increase the vehicle mass in Earth orbit, the spaceship can be used in subsequent missions, thus lowering their costs. An important part of the manned spaceship's configuration is how to protect astronauts against radiation hazards.[16] There are two types of radiation that need to be addressed for long-duration human space flight. The first originate from solar flares, which eject into space clouds of electrons, ions, and atoms via the Sun's corona. They produce streams of highly energetic particles in the solar wind, known as a solar proton event. The frequency of occurrence of solar flares varies, ranging from several per day when the Sun is "active" to less than one a week when it is "quiet." The second source is galactic cosmic rays, which, although less lethal than solar flares, constitute a continuous radiation to which the crew is exposed. Exact measurements of the cosmic-ray environment are yet needed to plan proper counter-measures.[17]

Several strategies are being studied to protect astronauts from the radiation hazard. One (passive absorbing shielding) is to build the spaceship with hydrogen-rich plastics such as a polyethylene-based material. Hydrogen is good for shielding against cosmic rays. The plastic materi-

[16] An instrument aboard the Curiosity Mars rover during its 253-day cruise to Mars calculated that the spaceship received around 660 millisieverts. The average yearly radiation dose received by a person on Earth is 3 millisieverts. It takes 250 millisieverts to cause any noticeable change in blood chemistry, and 750 millisieverts before any signs of illness occur. But at that level, recovery is likely; beyond that level, serious medical problems arise, so protection against radiation hazards on manned missions to Mars is a must.

[17] A physics experiment on the International Space Station will help in these measurements. The Alpha Magnetic Spectrometer will be attached outside the station and will search for different cosmic rays, measuring their long-term variation over a wide energy range, and for nuclei from protons of iron.

als' advantage is that they produce far less secondary radiation[18] than heavier materials such as aluminium or lead. Studies are progressing at NASA to develop polyethylene-based materials that combine superior structural properties with superior shielding properties. But shielding with hydrogen-rich plastic has a limit: if too much shielding material is used, this affects the efficiency of the mission. But a way out exists: limiting the mass of the shielding material can be achieved through the wise scheduling of the crew's activities, since significant amounts of crew time could be spent in an on-board sheltered area during the cruise phase. This limits the shielding-material mass while enhancing the safety of the crew.

A second strategy is to surround the spaceship with a magnetic field. Levy and French[19] first proposed this solution in the late 1960s. These early studies showed the intensity of magnetic field needed to deflect a proton with 2 Gev of energy, and the required electric-power input. New research has examined superconducting magnet technology to protect astronauts from radiation. These magnets are desirable because they can create intense magnetic fields with little electric-power input. With proper temperatures, they can maintain a stable magnetic field for long periods of time. The challenge is to keep the magnets at temperatures near absolute zero, where the superconductive properties of these materials emerge. But we are not yet there. Recent advances in technology and materials allow superconductive properties to exist at temperatures higher than 120 kelvin (–153 °C). In summary, superconducting technology has a low TRL, while passive absorbing technology is more advanced. TRL and safety considerations rule in favour of passive absorbing shielding and an on-board sheltered area of the spaceship.

A third strategy centres on increasing astronauts' resistance to the deleterious consequences of radiation exposure. A study commissioned by

[18] Secondary radiation comes from the shielding material itself. When particles of space radiation smash into atoms within the shield, they trigger tiny nuclear reactions, which produce a shower of nuclear by-products—neutrons and other particles—that then enter the spaceship.

[19] See R.H. Levy et al. (1964); and F.W. French et al. (1968).

ESA[20] is investigating the potential use by crew members of prophylactic radio-protective drugs and possibilities for controlling phytochemical antioxidants in the body through dietary choices. Scientists are looking for ways to help the body after the damage is done. As a cell divides, it pauses occasionally, to check its genes for any damage and to repair errors. With pharmaceuticals that lengthen this part of the cycle, researchers believe they can give the cell more of a chance to fix its own problems. Investigations of the biological response of the body to energetic particle radiation under conditions of micro-gravity are underway through dedicated animal space flight experiments.

4.1.6.2 Cargo and Descent/Ascent Vehicles

Cargo to be sent in advance to Mars orbit include: the habitat; power system and additional consumables; logistic components such as the long range traverse vehicle, emergency rover, and science equipment; and an ascent/descent vehicle. The latter is sent separately into Mars orbit. We support a configuration integrating the two vehicles. This solution allows the same propulsion system to be used for descent to, and ascent from, the surface of Mars. The number of landing vehicles depends on the payload sent in advance into Mars orbit. The landers' payload mass is estimated in Table 4.5. Four landers are needed according to the mission configuration.

In summary, the mission vehicles consist of five interplanetary spaceships. The first four are cargo spaceships. The descent/ascent vehicle, surface-habitat module, and tools for exploration are sent ahead of the human mission. The four cargo spaceships use chemical propulsion, reducing mission costs, and they could be upgraded versions of Ares V rockets, each of which is composed of three stages: an Earth departure stage, a trajectory insertion stage, and the lander. The fifth, heavier than the other four, is used to ferry the crew for the

---

[20] See Horneck, G. et al. (2003).

Table 4.5 Payload mass of Mars landers

	Mass (tonnes)
Habitat	**30**
Components of Adobe habitat and life-science module	10
Airlock with EVA suits	5
In situ resource utilisation equipment	15
Power	**35**
Habitat power system	30
Additional consumables	5
Logistics	**35**
Long-range traverse vehicles	20
Emergency rover	5
Science equipment	10
Ascent/descent vehicle	**46**
Cabin crew	6
Ascent stage with propellant	40

outbound and inbound trips. This manned spaceship is powered by a nuclear engine.

4.1.7 Landing Vehicles: Entry Descent and Landing Startegy

The descent/ascent module and other cargo modules are sent into Mars orbit before the human mission. When the manned spaceship arrives, the decent/ascent module will dock with it. The other cargo modules will instead make a direct entry to Mars and once they have landed the descent/ascent module will transport the crew to the surface. The thin atmosphere poses formidable challenges to safely landing the crew and cargo modules. Every mission undertaken by NASA relied on the same technology used in the 1976 Viking landing. This technology is based on a rigid blunt-body aero shell, a supersonic disk-gap-band parachute, and a subsonic propulsive descent system. Viking used these in sequence to decelerate the payload. Since then, engineers have worked with the same basic capabilities, incrementally tweaking them, using airbags to land the Pathfinder or using the elaborate sky crane to lower the Mars Science Laboratory (MSL) to the surface.

These missions delivered to the Martian surface payload masses from half a tonne to two tonnes W. This technology enables payload landings within 10 kilometres of the targeted landing site. With current technology nothing heavier than an MSL can be landed. Human exploration requires significant entry-method improvements, an increase in payload mass (30–40 tonnes), landing accuracy (in the order of metres), and the ability to land at higher altitudes (Table 4.6).

Various solutions are being researched. According to recent studies, rigid deployable heat shields or hypersonic inflatable atmospheric decelerators (HIADs) are promising systems. If the payload mass is of the order of 40 tonnes, HIADs appears to be more mass effective, but a large HIAD vehicle may have steering issues during the guided phase of flight. Rigid heat shields or smaller HIAD vehicles offer more guidance but at the expense of size or mass of the payload. The most promising technology for bringing large payloads to the surface of Mars is called supersonic retropropulsion (SRP), which involves starting the propulsive deceleration at supersonic speed by directing engine thrust into the oncoming free-stream flow. To fire rockets in the atmosphere while moving faster than the speed of sound is a hazardous undertaking. Turning on the propulsion at high speeds creates complicated shock waves, which can affect the attitude control[21] of the descent vehicle during the SRP entry phase and deceleration.

Table 4.6 Comparison of Mars Viking-based Entry, Descent, and Landing (EDL) systems

	Viking 1 & 2	Pathfinder	MER A & B	Phoenix	MSL
Aeroshell diameter (m)	3.5	2.65	2.65	2.65	4.5
Entry mass (t)	0.99	0.58	0.83	0.60	3.38
Ballistic coefficient (kg/m^2)	64	63	94	70	140
Relative entry velocity (km/s)	4.5	7.6	5.5	5.5	5.9
Hypersonic L/D	0.18	0	0	0	0.24
Parachute diameter (m)	16	12.5	14	11.7	21.5
Parachute deployment mach	1.1	1.57	1.77	1.65	2.1
Total landed mass (t)	0.590	0.360	0.539	0.364	1.7

[21] Attitude control is the control of the orientation of an object with respect to an inertial frame of reference or another entity.

Space agencies have no experience and little data on SRP and their effects on the descent vehicle. NASA has conducted a few tests inside a wind tunnel to analyse the effects of the jet's flow when fired into a supersonic stream. These experiments were part of the Constellation programme, but these were axed by a new administration. At present, NASA is carrying out thermal imaging and infrared-sensor data gathering studies of the controlled-descent tests of the SpaceX booster. The research team is interested in the 74 kilometre altitude range of the Falcon 9's re-entry burn, which is the relevant retropropulsion regime that models a Mars entry and the descent conditions. But much yet needs to be done.

Before beginning SRP flight demonstrations for Mars, multiple ground tests are needed to prove engine performance and provide data for computational fluid-dynamic (CFD) model validation. A series of Earth-based flight tests, scalable to Mars conditions, are needed to advance the technology to a level where the risks are reduced (Table 4.7).

4.1.8 Telecommunications

Support functions for telecommunications will monitor and control the human exploratory Mars mission. Telecommunications may be accomplished by direct links from the Martian surface to the Deep Space Network's antennas (DSN) on Earth or by relay links to the spaceship in orbit. A NASA DSN exists for communications with robotic probes sent to other planets and is a world-wide network with three main sites at the Goldstone Deep Space Communications Complex in California, the Madrid Deep Space Communication Complex in Spain, and the Canberra Deep Space Communication Complex in Australia. The three sites are located around the globe so that at least one of them can orient its antennas in a specific direction in space. Communication centres are linked with a control centre located in Pasadena. The network offers the vital two-way communications link that guides and controls interplanetary vehicles. Other countries have their own deep-space network. ESTRACK is the European space tracking network and Russia uses the Soviet Deep Space Network.

Table 4.7 Major technical challenges

Technology area	Major technical challenges
Propulsion	Developing large engines capable of throttling with sufficient thrust for human-scale payloads Demonstrating reliable engine start-up and throttling against supersonic flow
Aerodynamics/ aerothermodynamics	Understanding and predicting aerodynamics (static and dynamic forces and moments) and aerothermodynamics (surface heating)
Guidance, Navigation, and Control	Configuring the engines on candidate entry vehicle geometries to satisfy the required system performance Packaging the propulsion system within the volume and mass constraints of the system Verifying and validating the integrated system performance models (propulsion, flight mechanics, aerodynamics, aerothermodynamics, GN&C, thermal, structural) Developing entry-trajectory simulations using validated models
Ground testing	Testing in ground facilities to achieve relevant environments for engine, aerodynamics, and aerothermodynamic experiments Providing a database for validation of analytical methods (e.g. CFD)
Flight testing	Successfully executing stable and controlled instrumented flight tests at sufficient scale and complexity to satisfy TRL 5 and 6

Several factors make direct-to-Earth telecommunications challenging. These include the distance between Mars and Earth, the inability of Earth-based systems to communicate with all parts of Mars, and the power and mass constraints of the equipment on the Martian surface. Telecommunication through relay links provided by the spaceship in Mars orbit offers a more efficient solution with lower costs and decreases the communications range and provides coverage to all parts of the surface. The orbiting spaceship with telecommunications relay systems can send and receive large amounts of data to and from the Mars surface and relay these data to Earth. However, this solution presents several challenges: (i) incorporating highly autonomous operations for many of the local Mars telecommunications and navigation functions; (ii) achieving a high Mars-to-Earth data rate; and (iii) providing robust connectivity for manned links. But these challenges are not unsolvable and are being studied by several national space agencies.

4.1.9 Space Station Strategy

In the 1950s, a consensus existed on the orbital assembly of spaceships at a space station. However, this kind of assembly fell out of favour as rockets got bigger and technology enabled their better use. But this strategy was wrong then and remains wrong now. Orbital assembly's key virtue is to eliminate tight connections between expedition size and rockets. Smaller reusable rockets can be used to deliver the components of interplanetary spaceships into orbit to be assembled at the space station. Space agencies can accept deliveries from any company able to supply them. Competition will reduce costs. Moreover, space agencies do not need to fix the expedition's size when choosing the rocket. To estimate the size of an expedition from the start is difficult because it usually grows in size during the development phase. This is particularly relevant if the long-term goal is to build a permanent human base on Mars. In summary, a space port improves an interplanetary mission's logistics. Although costs can be high, investments may be justified by the lower costs of future missions.

In our scenario, the space station is built in phases between 2020 and 2040. If, as mentioned earlier, a habitable docking module is placed at a Moon–Earth Lagrangian point, it will form the space-station core. A first phase involves building the infrastructure for storage of minerals and fuel extracted from the asteroids by mining companies. This infrastructure would not be large. 3D printing, using minerals extracted from asteroids, could build many of the spaceships' components. The second phase involves adding to the core station the components for assembling and launching interplanetary spaceships[22] and for production and space tourism facilities. The new station's tasks would be the production of the

[22] This idea is not new. In April 2008, the Russian space agency proposed building an orbital construction yard (OPSEK) for spacecraft too heavy to launch from Earth. See http://www.parabolicarc.com/2009/06/29/roskosmos-administrator-perminov-speaks-present-future-iss-cooperation/#sthash.sKlb544Z.dpuf.

spaceships' components using metals from asteroid mining, the assembly of large spacecraft, flight tests and launches, and providing medical and biological facilities to rehabilitate interplanetary expedition crews following their return to lunar orbit. The following components should be added to the original module: one module for the in situ use of space mineral resources; a multi-task laboratory module, including biomedical facilities; and an assembling module for spaceships. If space tourism companies were interested, a module for a space hotel could be added.

The space station would be the single largest investment in space and could only be sustained through international cooperation, which might be organised through a non-profit company under the aegis of the United Nations. Membership should be open to all nations and private companies. Since the 1950s, the main actors in space exploration have been governmental agencies, enabling only privileged access to space. But the last decade has seen the rise of space tourism and ambitious private space mining companies, which require necessary space infrastructures to support their missions. Financing of the space station, both in terms of equity and loans, would then come from space agencies and private companies. Voting at meetings of the Board of Directors should be weighted according to its actual use by individual members and their investment quotas.

During construction, space agencies will guarantee access to the space station. Once it has been built, control of access should be left to private companies, freeing the agencies from this task. The space station can be financed by governments and private companies. Individual members' capital and long-term bonds guaranteed by governments will finance the project. A debt/capital ratio of 35/65 can be hypothesised. Revenues generated by the space station's operations will service the debt and are tolls levied against private enterprises and individual nations when they use it. Tolls should be calculated to compensate for the overcrowding and related costs created by its use.

4.2 The Costs of the Human Exploration of Mars

Estimated total costs of a manned mission to Mars range from a hundred billion dollars to a trillion dollars, the latter figure suggested by

President Bush when he unveiled a new space policy calling for such a mission. However, no detailed study has supported this number. The most advanced studies are those connected with NASA's Design Reference Mission (DRM). NASA issued a series of DRMs between 1993 and 2009. They are characterised by a small crew and several small interplanetary vehicles launched in one piece and larger interplanetary vehicles powered by chemical propulsion. The surface explorations, solely striving to achieve scientific goals, will be limited. A main constraint is the use of aero capture for Mars orbit insertion, so it is assumed that the size and shape of the interplanetary vehicles will be adequately configured to make such a capture safe and efficient.

The mission outlined earlier in this chapter differs in several respects from NASA's DRM. In the former the manned spaceship is powered by nuclear propulsion while the cargo spaceships are powered by chemical propulsion, as would be the case in the NASA missions. The scope for surface exploration of Mars is wider, and involves a heavier and more sophisticated long-range traverse vehicle. The habitat structures are manufactured using local resources. In contrast, the habitat in NASA's DRMs comprises pre-fabricated structures deployed on the surface. The base camp's power supply relies on a nuclear reactor, considered safer for long-stay missions. No space station is envisioned as part of NASA's DRMs. Cost estimates (Table 4.8) of the four landers are notional and draw heavily on the cost estimates performed by NASA and include research and development and construction costs. As significant revision should be expected after a thorough feasibility study, these costs should be treated with caution.

Based on the assumption that the four cargo vehicles are powered by chemical engines, the estimated costs of these vehicles are shown in Table 4.9. They include Ares V upgrade, construction, fuel, and Earth departure costs.

Table 4.8 Costs estimates of the landers ($ billions)

Ascent/descent vehicle	15
Habitat	5
Power	15
Logistics	15
Total	**50**

There are no estimates for developing and building an advanced nuclear-powered spaceship. Earlier we estimated that a nuclear-thermal spaceship's development costs would be around $3 billion, and we assume that the costs of an advanced nuclear spaceship would be multiples of this figure. The notional cost is set at $15 billion. For safety reasons, this spaceship could not be launched from Earth and would have to be assembled in space. We assume that the mass of the spaceship, including fuel, would be 150 tonnes. At least seven launches would be needed to bring the components to the space station using SLS rockets. An unofficial 2011 NASA document estimated the costs of the SLS programme through to 2025 to total at least $41 billion for four 70-tonne launches. This implies a cost of a 17 tonne payload launch of around $0.6 billion. The total cost of launching the components would then be around $10 billion. The total costs of building, launching, and assembling the spaceship would be around $25 billion. No estimates of costs are available for the space station. Based on ISS experience, this could be around $400 billion spread over 15 years and include the launching costs of materials and crew from Earth.

4.3 The Financing

In sum, the estimated total cost of the human exploratory mission to Mars, including the space station, is $475 billion. These costs are notional and subject to revisions. Governments and private companies would sustain the costs. Table 4.9 shows the costs divided between governments and private entities, though the assumption is that 80% of the space-station costs would be paid by governments. Financing will only be possible through international cooperation; and it is well known from the literature that the latter can leverage national investment on issues of

Table 4.9 Estimated costs of cargo vehicles

	Total costs ($ billions)
Habitat	3.0
Power	3.0
Logistics	3.0
Total	**9.0**

global relevance,[23] which improves affordability and returns value to every partner. In addition, cooperation is an opportunity for younger space agencies to take part in the exploration if mutual benefit for every partner is likely. Once cooperation is established, cancelling the programme becomes incompatible with political sustainability. Costs associated with the loss of expected benefits through the ending of such an international agreement would be greater than the costs of continuing it.

Space agencies and private companies will finance the space station through a special vehicle as outlined in paragraph 4.1.9. The space agencies' capital share is assumed to be 80 % (or around $208 billion). The private companies' capital share is assumed to be 20 %. Governments would guarantee long-term bonds ($140 billion) and space agencies would finance the exploratory mission. The total budget of the space agencies over a 20-year period would be $355 billion. On average, this implies a yearly increase of around $12.5 billion in space agencies' spending, or an average rate of increase of around 8 % per year. However, these outlays are bell-shaped with the peak rate of spending occurring at the time of construction of the space station and the spaceships.

A space-agencies' consortium under a United Nations aegis would carry out the exploration of Mars and would be open to every spacefaring nation. Investment quotas could be calculated according to the ratio of each space agency's budget within the total budget of participating agencies. The consortium would be responsible for designing the mission and coordinating the activities of space agencies and industries, which would allow the pre-programming of interoperability through the strategic use of redundancy. In case of default by any one of the chief partners, the maintenance of sufficient production capacity by the other chief partners would compensate for any lost skills relatively easily. Partners without critical technologies would reach agreements with chief partners should they not have adequate production capacity or when non-critical partners are able to supply technologies at lower costs.

---

[23] A good example is the European Organization for Nuclear Research (CERN), which operates the world's largest particle-physics laboratory. The large investment needed to build this laboratory induced 12 European nations to share the costs. The number of member countries subsequently grew to 28. CERN's main role is to provide the particle accelerators needed for high-energy physics research. As a result, many experiments at CERN involve international collaboration.

4.3.1 Finding Public Financial Resources

According to the 2014 Stockholm International Peace Research Institute (SIPRI) report, total military spending worldwide was $1.7 trillion in 2013. Global military expenditure increased on average by 14 % per year between 2000 and 2014, levelled off in 2012, and has been growing less rapidly both in 2013 and 2014. In the last two years, military spending fell in North America, Western and Central Europe, and Oceania, while it increased in other regions. Several factors explain these increases: real or perceived threats; the surge of terrorism; armed conflict; policies to contribute to multilateral peacekeeping operations; and the natural-resources curse.

The "natural-resources curse" is a phenomenon whereby nations find themselves in conflict due to power struggles over mineral resources, which induces the rapid growth of military spending by countries where the resources are produced. Sub-Saharan countries increased military spending by around 12 % per year between 2001 and 2012. The geopolitical competition for the control of strategic minerals contributed to the increases in military spending of China and Western countries. Assuming unchanged policies for the period of the next 20 years, cumulative military expenditures will be $32 trillion (Table 4.10). In our scenario, asteroid mining reduces international tensions and global military spending. Accumulated military expenditure might well decline by 5–10 % over a 20-year period, which translates to $1.6–3.2 trillion. To find $400 billion, spread over 20 years, among the resources freed up by the reduction in global military spending, should not be difficult. However, many factors will affect that decision.

One way politicians measure the benefits of large programmes is through the jobs and revenues generated within their constituencies. If the Apollo programme is taken as a reference, the jobs and revenue generated by Mars exploration will be an order of magnitude higher. Higher spending by space agencies offsets the negative impacts of short-term decreases in military spending. Higher spending benefits companies that work for the military-industrial sector, such as the aerospace industry (Table 4.2). Additionally, widening space exploration would have the support of both

Table 4.10 Expenditure by the military and space agencies ($ billions)

	2013	Extrapolation over next 20 years	%
Total military expenditure	**1557.3**	**32,354**	**100.0**
Military expenditure, USA	618.7	14,786	45.7
Military expenditure, other countries	742.6	17,568	54.3
Total space agency expenditure of which:	**73.7**	**1474.0**	**100.0**
NASA	43.6	872.0	59.2
Russia	2.7	54.0	3.7
ESA	5.6	112.0	7.6

Source: Military expenditures: SIPRI: "Yearbook 2014", SIPRI Publications, 2014

the companies and the people who work for them due to their direct interests in these programmes. Re-allocation of military spending should then be acceptable to any reasonable person (Table 4.11).

History suggests that civilisation moves in cycles. From the experiences of Babylon, Sumer, Egypt, Rome, and China, we know that civilisations eventually falter, due to natural disasters, corruption, social injustice, or constant wars. The classical world crises provide good examples of how those societies did not address the issue of justice, particularly that of slavery. Brilliant minds never challenged the political, economic, and religious foundations of their society. They apparently questioned everything in the fields of science, but never questioned slavery. The abundance of cheap slave labour likely was behind the classical world's inability to recognise the potential of technology to free people. Coupled with constant wars, this inability led to the demise of classical civilisations, which accomplished spectacular successes in science and an understanding of the natural world.

Presently we are experiencing an upward technological cycle, but events may cause this to decline. Awesome technological powers are concentrated in only a few people's hands, and few people representing the public interest seem capable of grasping this issue. Many have lost the ability to set their own agenda and/or to question those in authority. When many fall back, without noticing, into superstition and darkness, this is a prescription for disaster. We may survive this for a while, but

Table 4.11 Top ten weapon-producing companies

	Arms sales ($ billions)	% of total sales
Lockheed Martin (USA)	36.0	76
Boeing (USA)	27.6	34
BAE Systems (UK)	26.9	95
Raytheon (USA)	22.5	92
General Dynamics (USA)	20.9	66
Northrop Grumman (USA)	19.4	77
EADS (European Union)	15.4	21
United Technologies (USA)	13.5	22
Finmeccanica (Italy)	12.5	57
L-3 Communications (USA)	10.8	82

Source: SIPRI (2014)

Note: In the SIPRI statistics, Chinese arms manufacturers are not included, but it is believed that some of these groups may be ranked within the list of the top 20 companies producing weapons

ignorance and power may halt the upward technological cycle. If the human race becomes interplanetary, it may yet avoid this abeyance or make it as short as possible. It is thus wise to act when the window of opportunity is open rather than to rely on the possibility that it will stay open for a long time.

4.4 The Construction of a Permanent Base on Mars and Its Sustainability

If Mars exploration is accomplished by the late 2030s, a permanent base on Mars could be built in the 2040s. As the Cape colony if South Africa was a base that served ships venturing into the Indian and the Pacific oceans, so the Mars outpost will serve spaceships traveling between Earth and Mars and those beyond Mars. We call this outpost the "Base of Good Hope." Below we analyse the project's technical and economic feasibilities.

4.4.1 The Technical Feasibility

Several organisations are researching a permanent base[24] on Mars. The Mars Foundation is funding studies for the architectural design and for technologies related to the life-support systems and for equipment to build the base. Most studies concentrate on an underground permanent base, which would provide better protection against cosmic and solar radiation, wind, and dust. Besides radiation concerns, another issue with humans living on Mars is gravity. Humans have evolved in a 1 g gravity environment. Changing the magnitude of the gravity field to which the human body is accustomed is not wise, but 1 g gravity can be achieved by using a version of the gravity wheel,[25] assembled inside a doughnut-shaped cavity dug out of the underground base. Since the energy needed to sustain the base is large, a combination of nuclear, solar, and wind power can be jointly considered. Here the fundamental concept of "live off the land" is used to the greatest possible extent.

An important component of the base's architecture will be the greenhouses where agricultural commodities are produced. Plants need air, soil, water, and nutrients, all of which exist to some extent on Mars. The challenges are in making the elemental needs of plant life available. The first and most obvious problem in introducing human-sustaining agriculture to Mars is the nature of Martian regolith. Early studies of the soil show that it contains life-sustaining elements such as iron, potassium, magnesium, and sodium. Scientists as late as 2008 proclaimed the soil to be workable for plant growth. But that declaration was premature. Later, more thorough analysis of the soil derived from wider geographic areas led to the opposite conclusion. Martian regolith is now thought to be saturated with caustic superoxides such as potassium hydroxide. Ways to remove superoxides from the regolith are being developed.

---

[24] Principally the Mars Foundation. Other organisations are simulating technologies and equipment needed to build the Mars base. Simulations are conducted in zones with temperatures close to those on Mars, such as in the Antarctic and northern Canada.

[25] The idea of a rotating wheel to emulate Earth's gravity has been around since the start of the twentieth century. Wernher von Braun envisioned one version, where the wheel is changed so that the pods where people live are tilted downward. This enables the wheel to add more gravity to the existing Martian gravity. The downward angle of the pods and the wheel speed can be adjusted to provide a 1 g net gravity.

Another challenge is related to the lack of nitrogen and carbon dioxide in the Martian atmosphere. Overcoming this problem is not impossible. Carbon dioxide is part of the regolith, so releasing it from its bonds is only a technical issue. The regolith can then be used to produce enough gas for the growth of plants and to generate significant atmospheric pressure in greenhouses. A third problem is related to the lower gravity. While research on the ISS suggests plants can grow in microgravity, scientists do not know how the reduced gravity on Mars might affect different Earth crops. Further research on the ISS is needed. Finally, any pressurised underground greenhouse will block most of the sunlight reaching plants, so supplemental light is needed. These are not insurmountable problems, but a solution is needed before sending out the construction team.

A space elevator is also part of the base's architecture and consists of a cable with one end attached to the planetary surface and the other end in space. Stronger gravity at the lower end and outward/upward centrifugal force at the upper end result in a stationary cable being held up under tension. Space elevators reduce costs since loading and unloading to and from spaceships in orbit are mechanical operations. Their implementation on Earth is impossible with present technologies, because we have not yet the materials required to build a cable with enough compressive strength to support its own weight. Mars's low gravity, around a quarter of Earth's, enables building cables with existing materials.

As suggested by several researchers,[26] Mars's unique configuration and its nearby moon Phobos allows the construction of a simple system of space elevators, composed of one linking Phobos to Mars and another extending from Phobos outwards into space. The first elevator ends in the upper Martian atmosphere. Because Phobos is around 6000 kilometres above Mars's surface and well below geostationary orbit, materials to build the cable are available. The second elevator extends above Phobos so that a spaceship could dock at its end. Small craft will leave Mars's surface to conjoin with the first elevator which then will transport the materials to Phobos. From Phobos they are sent onward to the spaceship docked at the end of the second space elevator. Personnel and materials brought from the spaceship will follow the same route in reverse. This

[26] See L. M. Weistein (2003).

approach reduces the costs of transporting materials and personnel from the Martian surface to spaceships and the other way around.

It is hard to estimate precisely the time needed to build the base, but best guesses point to six months. The base is then ready to receive settlers, and the construction team can return to Earth or stay, depending on the effects of low gravity (a third that of the Earth) on the human body. These effects will determine if Mars's future residents will rotate at intervals or remain and initiate a new species: *Martian Homo sapiens.*

4.4.2 The Sustainability of the Base of Good Hope and Its Financing

Many factors influence the Base of Good Hope's sustainability. The base will not have for a long time the division of labour needed to make it self-sufficient. Initially, the population can be estimated between 50 and 100.[27] Once operational, the number of residents will increase to several hundred. The base will need to import from Earth manufactured goods and raw materials, including fertilisers or traditional manure to produce agricultural commodities, since Mars's soil is poor in nitrogen and other components necessary for agricultural production. The mass of these supplies will not be large and their prices on Earth are not exorbitant. But transport costs are high, which would increase their prices on Mars. Mars's current account balance could be brought into equilibrium by exporting precious raw materials from its moons and other degradable materials, servicing spaceships in terms of fuel and maintenance, and innovations.

We know from the United Nations Outer Space Treaty that no nation can claim sovereignty over celestial bodies. This excludes their colonisation (direct rule) by one spacefaring nation. The alternative is for private

[27] Who will be the future colonists? A first group will be individuals who place high value in individual freedom, personal control, and self-realisation. They will be the modern version of frontier men and women. A second group will be individuals who place high value on the search for new opportunities or simply desire adventure and exploration. They are the modern version of Homer's Ulysses. Today, as yesterday, personal values are diversified among the human population. A recent survey shows that the likely candidates constitute around 25 % of the population in advanced industrial countries; see Klinenberg E. (2012). The required skills will be mainly technical, geological, bio/geochemical, agronomical, and medical. Other skills, such as administrative, organisational, legal and logistics, should be assigned to individuals on the basis of their cross-training.

companies and individuals to seek dominion[28] status for the base under the aegis of the United Nations. This is similar, but not equal, to the institutional arrangements of the original Virginia colony.[29] Virginia started as a joint stock company, financed by individual investors hoping to make a profit. The colony was a business investment and a risky one. Later on, London revoked the Virginia Company's charter and transferred Virginia to royal authority as a Crown colony. The legislature of the Virginia colony was the House of Burgesses, which governed together with a colonial governor. This legislature was ready to establish British legal institutions even against the wishes of the home country.

An important consideration for the base to prosper is the institution's character. It is known from colonial history[30] that institutional characteristics play an important role in the economic development of colonies, including a fair system of property rights and representative government. Accountable governments providing investment incentives via stable policy making are the foundation for increased productivity and development. So the base should be granted autonomous governmental powers for internal management. The settlers' individualistic culture will help to set up a fair system of property rights, competition laws, and representative government. Their technological culture will help in creating institutions that ensure technological innovation, an important income source for the base. Property rights and other securities will ensure productivity and long-term economic growth. A United Nations treaty will define relationships with Earth, particularly for trade and security.

The base's legal status will help to finance investments. It is impossible to give a precise figure for investment costs, but they would be in the

[28] In English common law, dominions were autonomous political systems nominally under the British Crown. They had specific forms of fully responsible government (not to be confused with the nature of representative government). These institutions were able to legislate on every matter, except in the fields of foreign affairs, defence, and international trade. These powers remained under the control of the United Kingdom's Parliament.

[29] King James I granted a charter to the Virginia Company in 1606. He appointed the board of the corporation and endorsed the colony. Members of the English aristocracy who had wealth, power, and family connections were the leading members. In 1609, the king revised the charter and the colony's governing constitution. The London board appointed a governor, who selected members of the council governing the colony; see Craven W. F. (1957); Bemiss F. M. (1957).

[30] See D. Acemoglu et al. (2001).

range of $10 billion due to the investment made during the preliminary Mars exploratory mission. The space station and nuclear-powered spaceships already exist. New investments include equipment, landing vehicles, and power generation for the base. If these rough estimates are plausible, it would be possible to finance this project through a public–private partnership. Private capital and long-term loans from public financial institutions would finance investment for transportation and the base's construction. Given the profile of this venture risk, only public institutions could issue these debt instruments. Revenues generated by the base would repay long-term loans. The Martian government would then sell property rights to Martian soil to individual investors to generate initial revenue. The only land that exists on Mars is open, and there is an immense amount of it. It might seem worthless because it cannot be exploited right away, but it could be bought and sold if private property rights exist. For example, the Martian government could grant mining rights to private groups to survey land. Such claims would be tradeable on the basis of their future speculative worth. Trade with Earth would be enforceable by having the United Nations penalise any import from Mars made anywhere on Earth in defiance of the trade agreement, whose enforceability would make property claims on Mars secure.

4.5 Manned Exploration of the Solar System Beyond Mars

To go beyond Mars, the energy required by spaceships leaps enormously. Propulsion based on nuclear fusion would provide this. The physics of nuclear fusion is well known: two or more atomic nuclei collide at sufficient speed and energy to join to form a new atomic nucleus. In stars, each elementary fusion liberates energy significantly greater than the one generated by nuclear fission. Current research into fusion focuses on how to generate useful electric power, but the current designs for fusion reactors are inappropriate for space transportation: their mass is far too great. To build a compact fusion reactor to power spaceships poses great technological and scientific challenges. Nevertheless, NASA is studying several approaches.

The Fusion Driven Rocket (FDR), under development by NASA, is a revolutionary approach. The power source releases its energy directly into the propellant, not requiring conversion to electricity. It uses a solid lithium propellant that requires no significant tankage mass and is heated and accelerated to a high exhaust velocity. It has no significant physical interaction with the spaceship, thus avoiding damage to it and limiting both the thermal heat load and necessary radiator mass. It has high specific power (~1 kW/kg) at a reasonable mass scale (<100 tonnes). The key to achieving this stems from research on the magnetically driven implosion of metal foils onto a magnetised plasma target to achieve fusion conditions.[31] The energy from the fusion is thus utilised at very high efficiency. NASA believes that the FDR can be realised with little extrapolation from existing technology.

Another approach to obtain controlled fusion combines magnetic with inertial confinement. As in the magnetic method, fuel at low density is confined by magnetic fields while heated into a plasma. As in inertial confinement, fusion starts by compressing the plasma to increase its density and temperature, using plasma cannons (i.e. electromagnetic acceleration techniques) instead of powerful lasers. Thrust is provided by energy produced by nuclear fusion and a magnetic field that acts as a thrust plate. This propulsion has the advantage of smaller dimensions because of the efficiency of the double system; it also uses fuel more efficiently.[32] The Human Outer Planets Exploration group at NASA is studying this technology.

But fusion will not power a spacecraft any time soon. Proposals are only at the conceptual stage. Many nuclear-fusion engine's components have very low TRLs, and the efficiency of several energy forms released during fusion reaction is low. Only around 20 % of released energy is

---

[31] Several low-mass metal liners form a thick blanket surrounding the target plasmoid and compress it to fusion conditions. The radiant, neutron, and particle energy from the plasma is absorbed by the encapsulating metal blanket. This energy is adequate to vaporise and ionise the metal blanket. This hot, ionised metal propellant expands through a magnetically insulated nozzle producing high thrust at the optimal specific impulse.

[32] In this propulsion, only 20 % of the energy produced becomes kinetic energy. A significant fraction, around 70 %, is lost as electromagnetic radiation, mainly X-rays. This energy could be recovered, for example to make an auxiliary laser engine. In this way, exhaust velocities may reach between one-hundredth and one-tenth of the speed of light.

contained within the fusion particles' kinetic energy; 10 % is in the form of IR-UV radiation, and 70 % is released as X-rays. Few researchers have analysed how to reclaim X-ray energy by using auxiliary laser thrusters powered by waste X-rays. Laser-aided fusion-powered spaceships can achieve velocities of around 10 % of the speed of light with a mass ratio of 20. This velocity can be achieved if the efficiency of the laser X-ray is around 40 %.

4.5.1 Where to Go

If scientists and engineers apply themselves, nuclear-fusion-powered spaceships will be ready in the second half of the twenty-first century. Such spaceships are capable of high velocities between 1 % and 10 % of the speed of light. The asteroid belt is the first target, but it is a dangerous place. Courage and other human capabilities are fundamental for these missions. The 'old' human brain is hard to beat in solving multiple-choice problems such as weighing up the safety and attractiveness concerns from among different target objects. The Mars permanent base must supply fuel, food, and spare parts, as astronauts know that they cannot bring enough of them for the duration of the trip.

One highly desirable target will be 16 Psych,[33] which is one of the largest celestial bodies in the asteroid belt, with an effective diameter of 200 kilometres. Iron and nickel are expected to be primary metals on it, though radar observations show that water and volatiles are not present, making it an odd asteroid. Since the platinum group metals are associated with iron and nickel, this asteroid will be a target for future mining operations. While robotic missions have not visited it, a research group from the University of Arizona has proposed such a mission, whose goal is to place the probe into orbit to analyse better this unique celestial body. ESA is also considering a robotic mission.

Once tested in the asteroid belt, nuclear-fusion propulsion will power spaceships intended for the natural satellites of Jupiter and Saturn.

[33] Annibale de Gasparis discovered 16 Psych in 1852. Since then, it has been the target of several astronomical observations. Many believe that it has an exposed metallic core different from its bigger parent's body.

Among the Jovian satellites are smaller ones which were nearby objects captured by Jupiter's strong gravitational field. The other larger ones are the moons born from a protoplanetary disk surrounding Jupiter, which is a miniature version of the birth and structure of the solar system. A most interesting Jovian satellite is Europa, which has a mass similar to our Moon's, and has a ferrous metallic core and a magnetic field. It has a tenuous atmosphere, composed primarily of oxygen, and its surface is striated by cracks and streaks while craters are rare. On the surface, the water is frozen due to the low external temperature (–150°). Nobody knows the depth of the ice layer, which may vary from a few hundred metres to a few kilometres. The surface's smoothness and presence of frozen water has led to the hypothesis that a water ocean exists beneath the surface, heated by Jupiter's gravitational energy and underground volcanic activity. Since it is likely that oxygen is dissolved[34] in the subsurface ocean, this moon makes a case study for extraterrestrial life. Life forms could have started in Europa's depths which are like those found in the depths of Earth's oceans close to volcanic vents. A manned mission to Europa and drilling into its icy layers to reach liquid water may reveal fascinating results.

Saturn's satellite Titan is another high-interest celestial body. We know something of it following the landing of the Cassini–Huygens probe in 2005. It has a size similar to Mercury, which places it among the largest of the Saturnian moons after Ganymede. Titan is the only moon known to have a dense atmosphere, one composed mostly of nitrogen with clouds of methane and ethane. Weather, including wind and rain, creates surface characteristics similar to Earth. There are dunes, rivers, lakes, seas,[35] and deltas. Titan has seasonal weather, and a prebiotic environment may exist there today. The Cassini–Huygens mission did not produce evidence of complex organic compounds, but it encountered an environment similar

[34] The physical mechanism that makes possible the oxygen's presence starts from the dissociation of water molecules in the ice surface caused by the bombardment of cosmic rays. Oxygen molecules then spread into the liquid water through cracks in the ice caused by tidal motions.

[35] Data from Voyager 1 and 2 suggest the presence of hydrocarbons in Titan's seas and a thick atmosphere with approximately the correct temperature and to support them. Data from Hubble and other observations suggest the existence of liquid methane on Titan either in disconnected pockets or in wide oceans as on Earth.

to that theorised for the early Earth. Conditions on Titan will become far more habitable in the distant future. Five billion years from now, as the Sun becomes a red giant, Titan's surface temperature will rise enough to support liquid surface water. As the Sun's ultraviolet radiation decreases, haze in Titan's upper atmosphere will decrease. This will lessen the anti-greenhouse effect enabling atmospheric methane to play a far greater role. These conditions will create a habitable environment that could persist for several hundred million years.

We cannot imagine human exploratory missions beyond Saturn. Even with nuclear-fusion propulsion, these missions would require durations of ten years or more. Shorter human missions cannot be achieved with existing propulsion technologies. However, robotic missions will continue to Neptune and the Oort Cloud. Around Neptune, we find the mysterious moon Triton. It is a large satellite that orbits counter-clockwise to its planet. It may have originated as an object at the outer parts of the solar system, having been captured by Neptune's gravitational field. Intriguingly, Triton is one of the few moons in the solar system that is geologically active. Its surface is relatively young, with a complex geological history containing intricate and mysterious volcanic soils and tectonic activity. It has a tenuous nitrogen atmosphere with traces of carbon monoxide and methane near the surface. As with Pluto's atmosphere, Triton's might have originated by way of the evaporation of nitrogen from the surface.

Beyond Neptune, we find a huge population of comets in the Oort Cloud. Astronomers conjecture that matter composing the Oort Cloud formed closer to the Sun. It then scattered, early in the solar system's evolution, far into space because of the giant planets' gravitational effects. The outer Oort Cloud is loosely bound to the solar system and thus is easily affected by the gravitational pull both of passing stars and of the Milky Way itself. These forces occasionally dislodge comets from their orbits within the cloud and send them towards the inner solar system. Their water content and organic molecules have remained intact since the origin of the solar system around 5 billion years ago. Their closer analysis will supply a greater understanding of the origin of the solar system and life itself. At the Oort Cloud, we are near the edge of the solar system. Going beyond will be the subject of the next chapter.

4.6 Appendix: Various Propulsion Systems Proposed for a Manned Mars Mission

4.6.1 Chemical Propulsion

Several proposals[36] are under consideration for manned Mars missions based on liquid-propellant rockets. The most efficient fuel is liquid hydrogen, while combustion of hydrogen and oxygen is effective, which allows for a high specific impulse (around 450 s) compared to other propellant types, a technology that is already mastered. The Apollo missions used this technology for the trans-lunar injection. There are three challenges in using this cryogenic propulsion for a manned Mars mission:

- Keeping a spaceship's initial mass low. If used for the mission's primary propulsion phases, the total mass in low Earth orbit is around 1000 tonnes, assuming a crew of six astronauts. To reduce the initial mass, propellant for the return trip must be produced on Mars. Alternatively, fuel must be sent in advance and stationed in Mars orbit.
- Mastering the propellant's evaporation. Today we can store liquid hydrogen and oxygen in a spaceship for a few weeks. But several months of storage are needed for a Mars mission. Space agencies (NASA, CNES) are working to improve the tank's insulation and active cooling systems.
- Primary rockets are lost after every launch. If plans entail several missions to Mars, non-reusable rockets affect overall costs and are an inefficient solution.

In summary, chemical propulsion consumes large amounts of fuel and has low-cost efficiency. Other technologies are being looked at, such as nuclear fission that uses 50 % less propellant than the best chemical engine.

[36] See R. M. Zubrin et al. (1997); L. Bessone et al. (2004).

4.6.2 Nuclear Propulsion

Nuclear-fission propulsion is a broad topic that includes several engine designs, which range from the thermal-nuclear engine to nuclear-electric engines and more advanced designs such as the Rubbia engine.

Thermal-Nuclear Propulsion In thermal-nuclear propulsion, a nuclear reactor heats a coolant to high temperatures and expels it out through a nozzle. The most common type is the NERVA (Nuclear Engine for Rocket Vehicle Application), developed and tested in the 1960s. Their specific impulse is around 900 s, double that of chemical engines.

The main advantage of thermal-nuclear propulsion is its advanced stage of development (TRL). The United States and the then Soviet Union tested thermal-nuclear propulsion between 1960 and 1971. A few engines have operated for up to one hour with a thrust of 330 kN. Nuclear-thermal propulsion is still considered to be one of the key technologies by NASA and several related research activities within NASA and the DOE laboratories are ongoing. The Russians built and ground-tested several nuclear-thermal engines, which operated on a variety of propellants, including hydrogen, ammonia, and alcohol. Although their work continued until the mid-1980s, no evidence exists to confirm whether the Russians flight-tested any of these engines.

Nuclear-thermal propulsion offers other advantages: (i) shorter mission time (mission time to Mars is around 300 days, around half the time using chemical propulsion, due to the increased thrust); (ii) lower operating costs because of reduced propellant requirement; and (iii) efficiency of operations because the spaceship could be used in several missions to Mars, contrary to spaceships powered by chemicals.

The main challenges associated with the use of nuclear-thermal propulsion are:

- Radiation, which a nuclear engine gives off in large amounts. Before considering these engines, effective shielding mechanisms must be developed. The goal is to limit the crew's exposure to 10 REM. By comparison, the limit for civilians is 150 REM, and for military personnel

500 REM. The bad news is that this radiation shielding adds to the spaceship's mass.

- Testing a nuclear-thermal propulsion device is like testing a nuclear-power plant with the primary circuit open. Today, nuclear systems must have at least three confinement barriers: fuel cladding, a primary circuit envelope, and a confinement building for the reactor. Nuclear-thermal propulsion has a single confinement barrier: fuel cladding separating the uranium and fission products from the environment. To provide an environmentally safe system is thus a complex problem. Complexity could constitute a barrier to the development of these engines if not handled properly.

Nuclear-Electric Propulsion Nuclear-electric propulsion is composed of two main parts: a fission-based power-generation unit and an electric propulsion module. A variety of thrusters use the electricity generated by the nuclear reactor. Ion thrusters use electric fields to accelerate ions to high velocities. In principle, the only limit on the specific impulse achievable with ion thrusters is the operating voltage and the power supply. Hall thrusters use magnetic fields to ionise the propellant gas and to create a net axial-electric field accelerating ions in the thrust direction. MPD thrusters use either steady-state or pulsed-electromagnetic fields to accelerate plasma in the thrust direction. To get a high thrust density, ion thrusters use xenon, while Hall and MPD thrusters can work well with argon or hydrogen.

One advantage of electric-propulsion technology is its steady development since the 1950s. Ion and Hall-thruster technologies have matured to the point that they are now being used on commercial and military satellites with photo-voltaic power sources. Magnetic Plasma Dynamcs (MPD)-thruster technology is still under development. Another advantage is the higher specific impulses of the electric propulsion. In contrast to a chemical rocket or a nuclear-thermal rocket, they can work continuously for days, weeks, and even months. The main problem is that they have a low power-to-weight ratio and a low thrust density. This makes nuclear electric propulsion infeasible for missions where high accelerations are required. Nuclear-electric propulsion is well suited for unmanned cargo missions between the Earth, Moon, and the other planets.

Advanced Nuclear Propulsion Advanced nuclear propulsion is under study both in the United States and in Europe. The Idaho National Engineering Laboratory and the Lawrence Livermore National Laboratory advanced one proposal for a fission-fragment engine, which is an engine design that directly harnesses hot nuclear-fission products for thrust as opposed to using a separate fluid as the working mass. In a conventional nuclear reactor, the high kinetic energy of the fission fragments is dissipated by collisions to generate heat, which is then converted to electrical power with efficiencies of only 50%. Alternatively, the fission fragments produced in the plasma reactor can be used directly for providing thrust. This design can, in theory, produce high specific impulses, while still being well within the abilities of current technologies. In Fig. 4.2, **A** are the fission fragments ejected for propulsion; **B** is the reactor; **C** are the fission fragments decelerated for power generation; **d** is the moderator, either beryllium or lithium; **e** is the containment field generator; and **f** is the induction coil.

In this engine, fuel is placed into several very thin carbon bundles, each one sub-critical. Bundles are collected and arranged like spokes on a wheel, and several wheels are stacked on a common shaft to produce a single large cylinder. The entire cylinder rotates so that several bundles are always in the reactor core where the additional surrounding fuel make the bundles go critical. Fission fragments at the bundles' surface break free and are channelled for thrust. Lower-temperature unreacted fuel rotates out of the core to cool. The engine thus automatically selects the most energetic fuel to become the working mass. The engine's efficiency is high; specific impulses greater than 100,000 s are possible using existing materials. But the reactor-core's mass makes the overall performance of these engines lower.

In Europe, Rubbia designed an advanced nuclear engine between 1998 and 2002. This engine is powered by americium 242,[37] a fission-

[37] The comparative advantage of americium 242 over isotopes of uranium or plutonium lies in its chemical properties. Due to its lower critical mass, it can be produced as thin sheets of less than 1

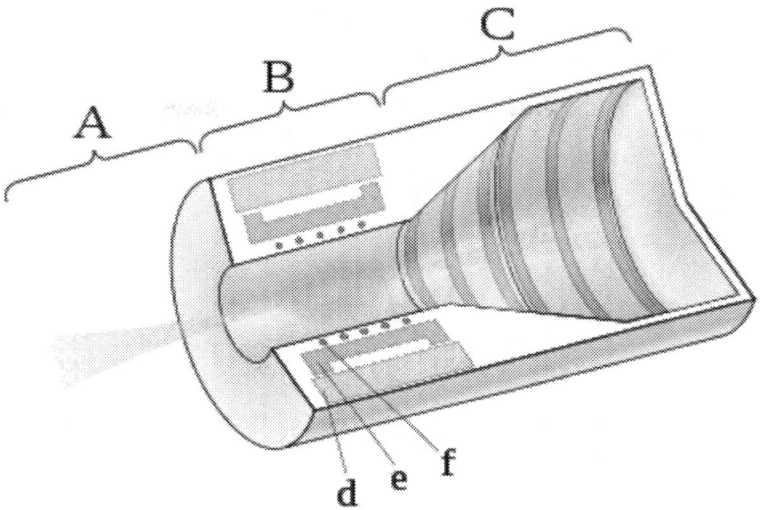

Fig. 4.2 Fission-fragment plasma-bed reactor (*Source*: Credit:http://en.wikipedia.org/wiki/File:Dusty_plasma_bed_reactor.svg)

able material hungry for neutrons and capable of absorbing them when bombarded by neutrons. The fissionable material is coated in a thin layer (less than a micron) on the inner walls of a hollow cylinder. This tube is filled with normal gas (hydrogen) at pressures of one atmosphere. Thanks to this geometry, when fission occurs, a high probability exists that one fission fragment flies into the gas and deposits its kinetic energy there. In the nozzle, a small miracle of physics happens. Inside the tube, the superheated gas moves in all directions with a chaotic motion. In the nozzle, the gas moves in one direction and pushes in the opposite direction. Newton is vindicated. The spaceship, anchored to the engine, moves in the opposite direction of the gas (Fig. 4.3).

micron. This enables fission products to ionise and escape the fuel easily. Carlo Rubbia and Giovanni Bignami proved that it takes the fission of a small quantity of americium 242 to produce a large amount of energy. Such ability to pack a big punch in a light and compact fuel allows for a smaller and lighter nuclear generator. Utilising such a generator in nuclear-thermal propulsion will allow for greater weight allocation to the payload, significantly improving the efficiency.

Temperature in the gas's central region can reach around 10,000 °C. With this heating method, a large saving of fissionable material is possible. The spaceship needs just a few pounds of plutonium or americium for a round trip to Mars. This mass is equivalent to two large cups (1 litre of plutonium weighs around 20 pounds). The specific impulse is estimated at 4000 s. Average exhaust speed is estimated at 40 kilometres per second. But a few unsolved problems remain. The first one concerns the material that can withstand the high temperatures of the super-heated gas. Research is ongoing to develop materials that can do this. The second problem concerns the dissipation of the heat generated, which can be done by equipping the spaceship with large heat-exchange surfaces facing the void of space. Because the temperature of space is around 2.7 degree above absolute zero, it has an infinite capacity to absorb heat.

In summary, nuclear propulsion is the best near-term method for powering spaceships directed at Mars. It offers lower costs and shorter travel times. Among possible nuclear engines, we have chosen the more advanced nuclear propulsion, called the Rubbia engine. It enables a shorter time of travel, less use of propellant, and more payload compared to other nuclear-propulsion systems. The major drawback of advanced propulsion is that they are today only at the concept stage. They need

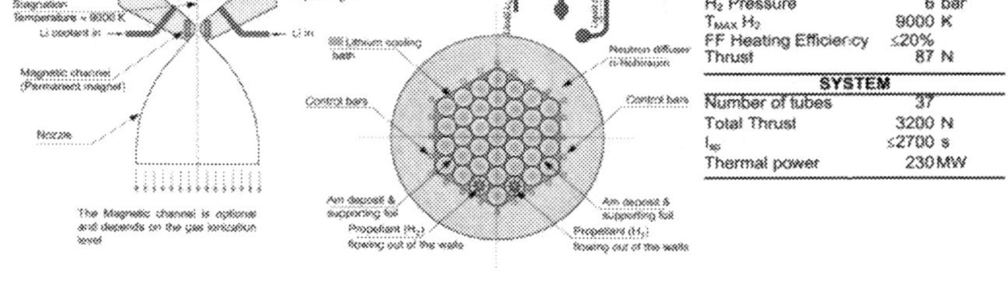

Fig. 4.3 Scheme of The Rubbia engine (*Source*: Bignami et al. 2011)

extensive engineering work and testing before spaceships powered by them are ready to fly. But our scenario foresees an exploratory Mars mission in the course of the 2040s. This timetable provides sufficient time for the development and testing of advanced nuclear engines.

References

Acemoglu D et al. The colonial origins of comparative development: an empirical investigation. Am Econ Rev. 2001;91(5):1369–401.

Bemiss FM. The three charters of the Virginia company of London. Virginia's 350th Anniversary Celebration Corp. Williamsburg, Virginia; 1957.

Bessone L et al. Human missions to Mars, overall architecture assessment. CDF Study Report CDF-20(A); 2004 Feb.

Bignami G et al. Project 242: fission fragments direct heating for space propulsion – program, synthesis and applications to space exploration. Astro Acta. 2011b;82:153–8.

Craven WF. The Virginia Company of London, 1606–1624. Charlottesville: University of Virginia Press; 1957.

French FW et al. Plasma radiation shield: concept and applications to space vehicles. J Spacecr. 1968;5(5):570–7.

George LE et al. Interplanetary mission design handbook: Earth to Mars mission opportunities and Mars to Earth return opportunities 2009–2024. NASA-TM-1998-208533; 1998.

Gugliotta G. The great human migration: why humans left their African homeland 80,000 years ago to colonize the world. Washington DC: Smithsonian Magazine; 2008.

Horneck G et al. HUMEX. A study on the survivability and adaptation of humans to long-duration exploratory missions. Noordwijik: ESA Publications Division; 2003.

Klinenberg E. Going solo: the extraordinary rise and surprising appeal of living alone. New York: Penguin; 2012.

Landau DF et al. Trajectories for human missions to Mars, part 1: impulsive transfers. J Spacecr Rocket. 2006;43(5). doi:10.2514/1.18995.

Levy RH et al. Plasma radiation shielding. AIAA J. 1964;2(10):1835–8.

Villanueva GL et al. Strong water isotope anomalies in the martian atmosphere: probing current and ancient reservoires. Science. 2015;348(6231):218–21.

Weistein LM. Space colonization using space elevators from phobos. Paper developed at the Advanced Measurement and Diagnostics Branch, NASA Langley Research Center, doi:10.1063/1.1541423; 2003.

Zubrin R et al. Mars direct: humans to the red planet by 1999. Proceedings of the 41st Congress of the International Astronautical Federation, Paris; 1990.

Zubrin RM et al. Mars direct: a simple, robust, and cost effective architecture for the space exploration initiative. Paper AAS 90–168. In: Meyer TR, editor. The case for Mars IV: the international exploration of Mars, part I, mission strategy & architectures. AAS Science and Technology Series Volume 89. Proceedings of The Case for Mars Conference; 1997.

5

The Ultimate Challenge: The Exploration and Colonisation of Extrasolar Planets

At the end of the twenty-first century, humanity will have explored out to our solar-system's boundaries, beyond which many have speculated for centuries about what may exist there. Plato and Aristotle, who dominated the cosmology of the ancient Greeks and Romans, considered the universe to be composed of the five planets visible to the naked eye, plus Earth and the Moon. What lay beyond was left to metaphysical speculation. However, Democritus claimed that countless stars composed the Milky Way and that they appeared as points of light due to their vast distances from Earth. Lucretius thought the universe to be infinite and populated by an infinite number of stars. The Catholic Church adopted Aristotle's mechanistic ideas (ipse dixit), which dominated the following centuries.

In the fifteenth century, Nicholas of Cusa revived the thinking of Democritus and Lucretius and argued that the universe was infinite and had a centre that was everywhere and a circumference that was nowhere. It contained an infinite number of stars. Contemporary astronomers did not take these insights seriously. Copernicus, Kepler, and Galileo continued to think that the universe's size was finite. But the Copernican revolution in astronomy put the Sun at the centre of everything. This gave the

G. Bignami, A. Sommariva, *The Future of Human Space Exploration*,
DOI 10.1057/978-1-137-52658-8_5

first, fatal blow to the Aristotelian anthropocentric view and to Catholic tradition. Only Giordano Bruno, a Dominican friar, philosopher, mathematician, and astronomer, argued that the universe contained an infinite number of worlds populated by other intelligent beings. This revolutionary idea could not be accepted by the Catholic Church that was having its own problems controlling the one Earth. For his cosmology, and for much more (including his declared atheism), Bruno suffered a tragic end.

With Newton, what lay beyond the solar system was removed from the realm of philosophical speculation and instead became part of science. But, despite the telescope's invention by Galileo and its subsequent improvements, the universe's size and how common solar systems might be remained unknown. We had to wait until the nineteenth and early twentieth century for answers to those questions. F. W. Bessel provided some preliminary answers in 1838. He employed the parallax method to measure the distance between Swan 61 and Earth. This was a great discovery, but with one important consequence: it is possible to use the parallax method only for a limited number of stars. For the majority, the angles involved in the calculations are very small and difficult to measure. The measurement problem received a strong impetus from Henrietta Leavitt's endeavours in 1890. She worked at the Astronomical Observatory of Harvard where she headed a group of women engaged in calculating the observations made by their male colleagues. Astronomers thought women were not suited to scientific work. But they could be employed in doing simple mechanical calculations, hence the name used to refer to them was "computers."

Leavitt developed a particular curiosity for stars called Cepheids. Information she received from the observers were measurements of the magnitude and periodicity of the brightness of these stars. She reasoned that this data concerned the stars' apparent brightness and not their absolute brightness. Studying the Cepheids in the Small Magellanic Cloud, she understood that these stars were at practically an equal distance from Earth due to the host nebula's enormous distance. This is what happens to an observer of a flock of birds. Although birds can be far from each other, they appear to be equally distant from the viewer's location. Based on this assumption, Leavitt realised that each Cepheid's apparent brightness indicated its absolute brightness relative to the other Cepheids. She

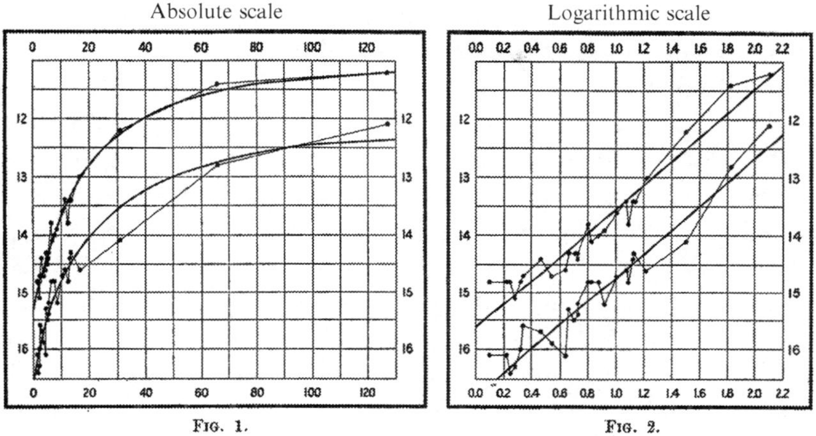

Fig. 5.1 Original Leavitt graph (*Source*: Leavitt and Pickering, *Harvard College Observatory Circular*, vol. 173, p.3, 1912)

then plotted a graph (Fig. 5.1) of the apparent brightness of 25 Cepheids versus their period of variation and discovered that the longer the period of variation the brighter the Cepheid.

She then formulated a mathematical equation for this relationship. It was then possible to compare any two Cepheids and to measure their relative distance from Earth. If one observes two Cepheids with similar periods of variation but with one being fainter than the other, the fainter one is farther away from Earth. Since brightness fades with the square of the distance, the mathematical formula enables the measurement of the star's distance from Earth. Leavitt published these findings in an obscure astronomy magazine[1] because women did not have access to the most important ones. Later, the Committee of the Nobel Prize for Physics became aware of her discovery, and they considered her for the Prize; but she died a few years later. Since the Nobel Prize can be awarded only to living scientists, she was denied this award which she rightly deserved.

Although Leavitt's method measures the relative distance of stars from Earth, she lacked a yardstick to measure absolute distances. If the

[1] See Leavitt H. S. (1908).

distance of one Cepheid from Earth were found, it would be possible to anchor Leavitt's measurement scale and to estimate every Cepheid's distance from Earth. A few years later, Herzsprung accomplished this. He determined the distances to several Cepheids by statistical parallax, thus establishing the first "standard candle" in astronomy. Herzsprung was thus able to calibrate the relationship discovered by Leavitt, which allowed astronomers to compute the distances from Earth to stars too remote for stellar parallax, so making it possible to measure the scale of the universe. Subsequently, Edwin Hubble discovered it was much bigger than astronomers imagined and that the Milky Way is just one galaxy among myriad others. But questions of how common stars with solar systems are and how similar their planets are to those in our own solar system remained unanswered.

5.1 The Discovery of Extrasolar Planets

The Observatoire de Haute-Provence made the first discovery of an extrasolar planet[2] in 1995 by measuring a gravitational influence on the parent star's motion. They used the Doppler effect to analyse the motions and properties of the star and the planet, both of which orbit a common centre of mass. They gravitationally attract each other, causing the star to wiggle around this common centre. As the star approaches the observer, the light emitted shifts to higher frequencies (blue shift); as it recedes, the light shifts to lower frequencies (red shift). The Doppler technique provides information on the star's velocity toward or away from Earth (star radial velocity). The planet's velocity can then be derived from the star's radial velocity. Once the star's and the planet's velocities and the star's mass are known, the planet's mass can be calculated. This method has the advantage of being applicable to stars with a wide range of characteristics, though it cannot measure the planet's true mass, as it can only set a lower limit to this because the angle of the orbital plane is not known. Another disadvantage is that small, Earth-like, extrasolar planets are harder to detect than giant massive planets.

---

[2] See Mayor M. et al. (1995).

From 1995 onwards, the detection of extrasolar planets has multiplied. Astronomers now use different methods to detect them. A first method (the transit method) consists of observing variations in the star's apparent luminosity as a planet crosses it. This variation is equal to the ratio of the planetary and stellar disc areas. The planet's size is ascertained from the star's luminosity variation and size. The star's orbital size and temperature enable us to calculate the planet's characteristic temperature,[3] which is a gauge of its habitability. As a rule of thumb, one may define the "habitable zone" as one where liquid water exists on the planet's surface. Liquid water is important for life to exist.[4] Moreover, the transit method makes possible the study of the planet's atmosphere and its composition.[5] Atmospheric composition is a significant indicator of a planet's habitability. Oxygen and nitrogen are important chemical elements for life, the former being unstable and whose presence in the planet's atmosphere is a good indicator of the presence of external agents keeping the atmosphere in disequilibrium. We know that only the process of photosynthesis, performed by plants or cyanobacteria, creates the conditions for disequilibrium in Earth's atmosphere. But we cannot rule out other unknown agents in extrasolar planets. The transit method has two major disadvantages. First, planetary transits are only observable for planets whose orbits are aligned with Earth; but less than 1 % of stars have this kind of orbit. If the angle is different, the planet never appears to pass in front of its star. Second, the method suffers from a high rate of false detections. A 2012 study[6] found that the rate of false detections for transits, observed by the Kepler mission, were as high as 40 % in a single planetary system.

---

[3] A secondary eclipse enables the measurement of the planet's radiation. If the star's photometric intensity during the secondary eclipse is subtracted from its intensity before or after, only the signal caused by the planet remains. It is then possible to measure the planet's temperature.

[4] Water is involved in all body processes. It is an efficient heat conductor and serves to maintain uniform body temperature. Water is indispensable as a protector of internal organs, serving as a cushion and preventing the transmission of outside shocks. It serves as nature's solvent for many chemical compounds and is the medium for many chemical reactions.

[5] When a planet transits the star, the star's light passes through the upper atmosphere of the planet. The planetary atmosphere can be detected by measuring the starlight's polarisation as it passes through or is reflected off the planet's atmosphere. By studying high-resolution stellar spectra, one can detect elements present in the planet's atmosphere.

[6] See Santerne A. et al. (2012).

A second method is called the transit-time variation, which enables us to detect more planets even if they do not transit. When multiple planets are present, each one perturbs the others' orbits. Small variations in the transit time of a planet show another planet's presence. NASA's Kepler satellite made the first significant detection of a non-transiting planet.[7]

A third method is astrometry, which measures a star's position in the sky and observes changes in that position that are due to a planet's gravitational influence. Astrometry works best when the planet's orbit around its star is perpendicular to the observer. If not, a shift in the star's position cannot be measured. As changes in a star's position are small, this method has produced few detections, but it has been used to investigate the properties of planets found in other ways.

A fifth method is gravitational microlensing, which exploits one of the predictions of Einstein's theory of general relativity; that is light follows the curvature of space–time. The star mass deflects light because it warps space–time. Gravitational microlensing occurs when a star's gravitational field acts as a lens deflecting the light of a distant background star. If the foreground star has a planet, then that planet's own gravitational field causes an asymmetrical perturbation in the signal. This effect occurs when the two stars are aligned. Since microlensing observations do not rely on radiation received from the lens object, this effect enables the study of massive objects no matter how faint they are. The microlensing effect is wavelength-independent, enabling the study of source objects that emit any electromagnetic radiation. It is thus an ideal technique to study the galactic population of such faint or dark objects as brown dwarfs, red dwarfs, planets, white dwarfs, neutron stars, and black holes. Lensing events are brief, lasting for weeks or days, as the two stars and Earth are moving relative to each other. Since this requires an improbable alignment, many distant stars must be continuously monitored to detect extrasolar planets. This method is most fruitful for planets between Earth and the centre of the galaxy as the galactic centre provides many background stars.

[7] The transiting planet Kepler-19b exhibits transit-time variations with five minutes of amplitude and a period of around 300 days. This suggests the existence of another planet, Kepler-19c, whose period is close to a rational multiple of the transiting planet's period.

In October of 2014, the Extrasolar Planets Encyclopedia[8] reported 1332 confirmed extrasolar planets. This count includes 1115 planetary systems, of which 459 contain more than one planet that has been detected. Most planets have large masses (Jupiter-like) and orbit near their stars. Few planets have masses similar to Earth, and, among these, few orbit their sun in the so-called habitable zone. Over the next decade, space-based telescopes, such as the NASA Kepler[9] and the James Webb Space[10] Telescope, will enable the discovery of more Earth-like planets and analysis of their atmospheres. Although the distance to its star is a significant indicator of a planet's habitability, other factors are important as well. One is the presence of an atmosphere containing chemical elements, such as oxygen, which will be necessary to sustain future colonisation of the planet. Another important factor is the presence of a moon. On Earth, the presence of our Moon has a powerful stabilising influence. The Moon's gravitational influence allows Earth's obliquity to oscillate by about 1.3 degrees about its mean position of 23.3 degrees. If the Moon did not exist, Earth's obliquity would evolve chaotically, similarly to what happened on Mars. Chaotic evolution of obliquity would have had dire consequence for the climate on Earth and complex life would not have had time to evolve, hence the importance of the presence of a moon around a planet whose habitability had been determined by its distance to a star.

Finally, the planet should have a magnetic field around it, since an absence would leave the planet at the mercy of fast moving electrically charged particles that are blown towards the planet by its star. This would

[8] See http://exoplanet.eu/catalog/.

[9] Kepler's sole instrument is a photometer that monitors the brightness of over 145,000 main-sequence stars. The photometer points to a field in the northern constellations of Cygnus, Lycra, and Draco. This is well out of the ecliptic plane, so that sunlight never enters the photometer as Kepler orbits the Sun. The stars observed by Kepler are roughly at the same distance from the galactic centre as the solar system and are close to the galactic plane.

[10] The James Webb Space Telescope (JWS) is a planned space observatory scheduled to be launched in October 2018 and will offer unprecedented resolution and sensitivity to radiation from the long-wavelength to infra-red. The project is an international collaboration of about 17 countries led by NASA. Its primary scientific mission has three main components: to search for light from the first stars and galaxies in the universe; to study the formation and evolution of galaxies; and to understand how stars and planetary systems formed. JWS will detect more extrasolar planets and allow the study of their atmospheres.

gradually blow away any atmosphere around the planet, leaving in the end almost nothing, as happened to Mars. Although the planet may be habitable in the short term, in the longer term it would become difficult to inhabit by human beings. In the near future, both terrestrial and space-based telescopes will allow the analysing of the spectrum of the light of planets to ascertain the presence of other chemical elements, such as oxygen. However, with present technologies, it is not possible to detect the presence of a moon around a planet orbiting its star in the habitable zone, nor can the presence of a magnetic field around it be detected. Only an exploratory probe would be able to gather this information, which would confirm that the planet is suitable for human colonisation.

5.2 How to Get There: Einstein and the Theory of Special Relativity

Interstellar distances are large. The closest stars to Earth lie in the triple star system Alpha Centauri, which is around 4.5 light years from Earth. To reach these stars in reasonable time, it would be necessary to achieve velocities equal to significant fractions of the speed of light. To attain such speed and lose it when reaching a destination will pose formidable challenges to spaceship designers. When approaching the speed of light, the spaceship must use more energy to achieve the same acceleration as at lower speeds. In theory, nuclear-fusion propulsion can run with continuous acceleration and achieve significant fractions of the speed of light, but the mass of fuel required is large. Without burying the reader in complicated calculations, the mass of fuel should be a multiple of the spaceship's mass. The conclusion is that interstellar travel with nuclear-fusion propulsion, although possible, is not practical. Bussard[11] suggested a way out, namely, to collect more fuel during the trip. Hydrogen is the most abundant material in the universe, and the average density of hydrogen in our galaxy is one hydrogen atom per cubic centimetre. To gather one gram of hydrogen per second, the spaceship could use collection panels more than 40 kilometres wide; another impractical solution.

[11] See Bussard R.W. (1960).

To go faster, we know the recipe: more efficient fuel with a higher energy content per unit mass. The annihilation of particles with their antimatter counterparts has the highest ratio known in physics between energy produced and mass used. When matter and antimatter meet, both "disappear," resulting in energy according to Einstein's equation $E = mc^2$. Multiplication by such a large number as the square of the speed of light shows that a small mass may give rise to a great deal of energy. Articles in the technical literature[12] have analysed antimatter as an energy source for interstellar vehicles. The only engine that can reach 50–60 % of the speed of light is one that uses the products of matter–antimatter annihilation directly to generate thrust. This system is called the pion engine. Pions are produced by proton–antiproton annihilation. The resulting charged pions will have a velocity of 94 % of the speed of light. A Lorentz factor of 2.93 extends a pion's lifespan sufficiently to travel 2.6 metres through the nozzle before decaying into muons. Sixty per cent of pions will have either a negative or a positive electric charge; 40 % of pions will be neutral. Neutral pions decay into gamma rays, which are useless for generating thrust. In realistic matter–antimatter reactions, collimation may not be perfect, and few pions are deflected backwards by the nozzle. Thus, the effective exhaust velocity for the entire reaction drops to between 50 and 60 % of the speed of light.

The antimatter mass needed to power engines for interstellar travel is large, ranging from a few kilograms to a few tonnes, depending on spaceship mass and cargo. There are two problems, however: the production and the storage of antimatter. Antimatter does not exist in the observable universe;[13] we have to produce it. So one main problem is how to produce antimatter and at what cost. At CERN in Geneva, particle accelerators produce 10^{15} antiprotons per year. If this seems like a large number, consider that to make one gram of antimatter you need 10^{26} antiprotons. At

[12] See Lewis, R.A., et al. (1997); and Frisbee R. (2003).

[13] Signs of the presence of antimatter in the universe come through the annihilation of matter and antimatter. If these two particles coexist, we can expect to see characteristic radiation emitted by annihilation, but no sign of this has been detected. The radiation that is detected is generated by the collision of high-energy cosmic rays with each other or with Earth's atmosphere. Small antimatter amounts are produced in these collisions. These results show that antimatter exists in a secondary form and in small amounts in the universe.

this rate, the production of one gram would take centuries, not to mention what would be required to produce kilograms or tonnes. Moreover, to produce antiprotons needs energy, but unfortunately, the energy needed is greater than that obtained by converting mass into energy. This raises the cost of producing one gram of antimatter to astronomical levels. G. R. Schmidt[14] has proposed a way to calculate the cost of antimatter production: converting input energy (E_{in}) into the energy (E_{out}) of the collected antiprotons' mass can be expressed in terms of efficiency:

$$\eta = \frac{E_{out}}{E_{in}} \tag{5.1}$$

Because of the equivalence between E_{out} and the antiproton mass M_a ($E_{out} = M_a \cdot c^2$), the energy required to create a unit mass of antiproton is given by:

$$\frac{E_{in}}{M_a} = \frac{c^2}{\eta} \tag{5.2}$$

The energy needed to produce antiprotons is inversely proportional to the conversion efficiency factor (η) as the speed of light is constant. The total cost of energy (C) to produce antiprotons is then obtained by multiplying the Eq. (5.2) by the grid energy cost (C_{grid}) and by M_a:

$$C = \frac{C_{grid} \cdot M_a \cdot c^2}{\eta} \tag{5.3}$$

In present particle accelerators, η is around 10^{-08}. Assuming C_{grid} to be around \$0.1 per kilowatt hour, this translates into an energy cost of around \$63 trillion per gram of antiprotons. So we are still far from producing a cost-efficient fuel to power a matter–antimatter engine. If the cost of antimatter is brought down to \$10 million per milligram, matter–

[14] See Schmidt G. R. and al. (1998).

antimatter propulsion will be competitive with nuclear-fusion propulsion, depending on the efficiency of the design of the various engines.

In the half century since the antiproton discovery, the intensity of antiprotons beams has increased by at least a factor of 10^4. This increase was not linear but exponential and is well documented. Who knows what accelerator physicists will be able to do to increase the efficiency of the production of antiprotons in the next half century. Current particle accelerators produce antiprotons only for scientific experiments, which is not the most efficient way. Few visionaries conceive the construction of accelerators dedicated to the production of antiprotons, which would decrease power losses during production and increase efficiency. Robert Forward, however, is one such visionary and has[15] analysed what happens when a particle accelerator is designed for no other purpose than to create antiprotons. Results show energy efficiency increases from one part in sixty million to one part in 10,000. This important improvement approaches the "break-even" point compared to nuclear-fusion engines.

Meanwhile, the costs of building particle accelerators can be lowered. Scientists are working at the SLAC National Accelerator Laboratory to reduce the size and cost of particle accelerators. The money invested to upgrade present facilities is estimated at between \$20 and 30 billion, but this will not come out of space agencies' budgets. It could come out of the budgets of the centres producing the antimatter once they realise the potential demand, which may increase not only by the prospective demand for interstellar exploration but for medical and other related uses. If prices fell, antimatter may find new applications in other areas. These developments may reduce the cost of antiprotons to \$10 million per milligram.

Once efficient ways to produce antimatter are found, the remaining problem is to conserve antiprotons. At Fermilab and CERN, they do it with special magnetic bottles known as Penning traps.[16] Results

[15] See Forward R. (1985).

[16] In a vacuum, the particles of antimatter can be trapped and cooled within a magnetic trap using a highly concentrated laser beam. Another method, more hypothetical, is the storage of antiprotons inside fullerenes. In theory, the negatively charged antiprotons can repel the electron cloud around the nucleus of carbon. Antiprotons do not get close enough to the protons of the nuclei and so avoid annihilation.

have been modest: experiments show that one billion antiprotons can be stored for one year. This is far from the antimatter mass needed by interstellar spaceships, but it is a start. A century ago few visionaries[17] imagined using liquid propellant for rockets. Even though it was a big step forward from solid propellant, scientists and many commentators mocked them. They claimed that producing liquid hydrogen and using it in an engine was too costly and dangerous. Half a century later, the use of liquid hydrogen had become a reality and its costs were reduced. More than half a century after the first micro-droplets of liquid hydrogen were produced in a Dutch laboratory, it was used to send men to the Moon.

The methods described above make it possible to produce small quantities of antimatter at a reasonable cost. Spaceships powered by matter–antimatter propulsion could then be built. The engines would use energy from the annihilation to heat a propellant for thrust.[18] The main differences between these engines are in their performance characteristics: their exhaust velocity, energy efficiency, and mass ratios. For example, using ACMF propulsion, the exhaust velocity can reach 100 kilometres per second. Using AIM propulsion, the exhaust velocity can reach 1000 kilometres per second. The antimatter needed for a one-year round trip to Jupiter by a spaceship powered by an ACMF engine is around 10 μg. Antimatter needed by a spaceship powered by an AIM engine for a 50-year round trip to the Oort Cloud is around 100 μg.

Antimatter needed for interstellar travel will be instead in the order of kilograms or tonnes. To produce these amounts requires technologies that go beyond those already described. Again, Forward in his book

[17] The idea of a liquid rocket as understood in the modern context first appeared in the book *The Exploration of Cosmic Space by Means of Reaction Devices* by K. Tsiolkovsky. During the nineteenth century, the only known developer of liquid-propellant rocket engine experiments was the Peruvian scientist Pedro Paulet.

[18] A proposed engine is called the Antimatter-Catalyzed Micro-Fission/Fusion (ACMF). In this engine, a pellet of deuterium-tritium and uranium 238 is compressed by particle beams and irradiated with a low-intensity beam of antiprotons, which are absorbed by the uranium 238. This initiates fission that heats and ignites the deuterium-tritium core, which expands and produces the pulsed thrust. A second proposed engine is called the Antimatter-Initiated Microfusion (AIM). In this engine, antiproton plasma is compressed by electric and magnetic fields, while droplets of deuterium-tritium are injected into the plasma. The antiprotons annihilate on contact with the deuterium-tritium droplets, which heat the plasma to ignition conditions. The products of this ignition are directed to a magnetic nozzle to produce thrust.

Indistinguishable from Magic[19] anticipated a possible solution. He proposes to move antimatter production from Earth to space:

> Where can we get the energy needed for the large-scale production of antimatter? Some of the prototypes will be built on Earth, but for the large scale production we do not want to feed these machines with the combustion of fossil fuels on Earth. There is plenty of energy in space. At the distance of the Earth from the Sun, the Sun provides more than a kilowatt of energy per square meter of collector or a gigawatt per square kilometre. A collector field of 100 square km would provide a power of ten terawatt, which is sufficient to produce one gram of antimatter per day.

Remember that in half a century asteroid mining will be in full swing. Raw materials from space will be available to manufacture space infrastructure at reduced costs. It is thus plausible to produce antimatter in space at reasonable costs, thus opening the way to interstellar travel.

5.3 Where Antimatter Propulsion Will Take Humanity

In 50–100 years, if we succeed at producing antimatter in large quantities at reasonable costs, the first interstellar mission could be organised. A spaceship powered by matter–antimatter could travel at most at 60 % of the speed of light. This restricts the mission's targets to within 10–15 light years from Earth (Fig. 5.2). At the moment, astronomers have identified six prime targets within 15 light years (Table 5.1). Alpha Centauri B is the closet star (4.3 light years away), followed by Barnard Star, Epsilon Eridani, Tau Ceti, Kapteyn's Star, and Gliese 832. These stars have planets whose orbits place them in the habitable zone.

Alpha Centauri is slightly older than our solar system: estimates range from 4.5 to 7 billion years. It is thought to have formed with elements heavier than hydrogen and helium, hence there is a high probability of the existence of rocky planets. Astronomers have found one such planet

[19] See Forward R. (1995).

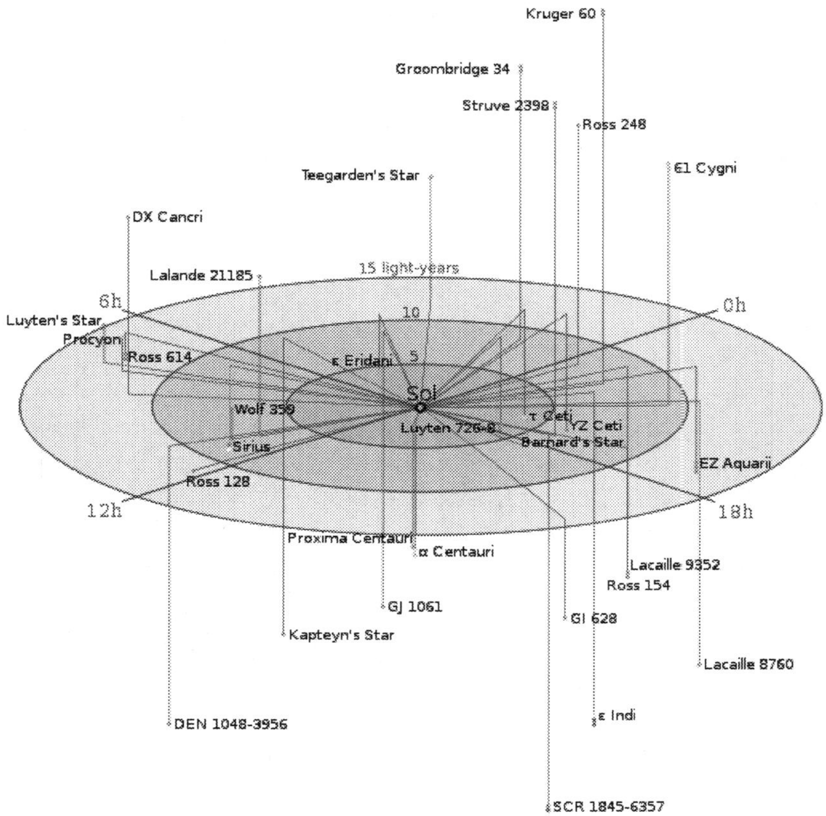

Fig. 5.2 Closest star systems to Earth (*Source*: https://en.wikipedia.org/wiki/List_of_nearest_stars_and_brown_dwarfs#/media/File:Nearby_Stars_(14ly_Radius).svg)

orbiting Alpha Centauri B, but it is too close to its sun to be habitable. The same astronomers argue that there may be other planets whose orbits are in the habitable zone. These hypothetical companions are difficult to find with current instruments. At present, spectrometers can detect changes in radial velocity of around 30 centimetres per second. To find planets with a mass similar to Earth, spectrometers would need to measure changes in the radial velocity of around nine centimetres per second. We need to build new, more powerful spectrometers.

Table 5.1 Possible targets for interstellar missions

Stellar system	Distance from Earth in light years	Remarks
Alpha Centauri	4.3	Triple system (G0, K5, M5). Component almost identical to Sun. High probability of "life bearing" planets
Barnard's Star	6	Very small, low luminosity red dwarf (M5). Closest system known to have one, and perhaps two or more planetary companions
Lalande 21185	8.2	Red dwarf star (M2) known to have a planet
Epsilon Eridani	10.8	Single star system; slightly smaller and cooler than the Sun (K2), may have a planetary system similar to our solar system
Tau Ceti	11.8	Single star system, similar in size and luminosity to the Sun (G4). High probability of possessing a "solar-like" planetary system
Kapteyn's Star	12.6	Red dwarf star (M2) known to have a planet in the habitable zone, Kapteyn b. One other planet has been detected
Gliese 832	15	Red dwarf star with confirmed planetary system composed of four planets

Barnard's Star is a low-mass red dwarf around six light years from Earth. At seven to twelve billion years of age, it is older than the Sun and might be among the oldest stars in the Milky Way galaxy. In 1969, Peter van de Kamp[20] claimed detection of one or more planets with masses comparable to Jupiter. For a decade, astronomers accepted this claim, but two important papers in 1973 contested it. Null results for planetary companions continued throughout the 1980s and 1990s, and observations based on interferometry from the Hubble Space telescope in 1999 were the latest. While research has restricted the possibility of the existence of planets around Barnard's Star, it has not ruled them out.

Lalande 21185 is a red dwarf star in the constellation of Ursa Major. Although relatively close by, it is only magnitude 7 in visible light and thus is too dim to be seen with the unaided eye. At approximately 8.31

[20] He had detected a perturbation in Barnard's Star's proper motion consistent with having one or more planets comparable in mass to Jupiter. See Van de Kamp P. (1969).

light years away from Earth, this star is the fourth closest stellar system to the Sun; only the Alpha Centauri system, Barnard's Star and Wolf 359 are known to be closer. In 1996, George Gatewood announced the discovery of multiple planets in this system, detected by astrometry. He claimed that such planets would usually appear more than 0.8 arc seconds from the M dwarf itself. However, subsequent searches by others, using coronagraphs and multifilter techniques to reduce the scattered-light problems from the star, have yet to identify positively any such planets.

Epsilon Eridani is at a distance of 10.5 light years from Earth and has an apparent magnitude[21] of 3.73. Its age is estimated at less than a billion years and because of its youth it has a higher level of magnetic activity than the present-day Sun. It is a target for planet finding programmes because it has properties allowing an Earth-like planet to form. Astronomers have observed its motion along the line of sight to Earth for over 20 years. Periodic changes in this data have yielded evidence of a giant planet orbiting it with a period of around seven years. It has a mean separation of 3.4 astronomical units.[22] The proximity of a large planet to Epsilon Eridani's habitable zone reduces the likelihood of finding a terrestrial planet with a stable orbit within it.

Tau Ceti is a star in the constellation Cetus. It is spectrally like the Sun and its mass is lower (78 %) than the Sun's. It is at a distance of just under 12 light-years from Earth and is stable with little variation. The principal interest in Tau Ceti are its Sun-like characteristics and their implications for possible planets and life. It was the target of a few radial velocity planetary searches, which found no periodical variations. The velocity precision reached so far is around 11 metres per second. This result excludes the presence of hot Jupiters, but Earth-like planets are not excluded. Since December 2012, observations have detected five planets in orbit. According to this study,[23] these planets have masses from two to six times Earth's; two are in the habitable zone.

[21] A star's apparent magnitude measures its brightness as seen by an observer on Earth adjusted by the value it would have in the absence of any atmosphere.

[22] See Artie P. et al. (2010).

[23] See Astrobiology Magazine, "*Tau Ceti may have a habitable planet*", December 19, 2012.

Kapteyn is a class M1 red dwarf star[24] around 12.6 light years from Earth in the southern constellation Pictor. The Dutch astronomer, Jacobus Kapteyn, originally catalogued it in 1898. The star's mass is between one-quarter and one-third of the Sun's, but is much cooler at around 3500 K. On 3 June 2014, a team of astronomers[25] reported the discovery of two super-Earths orbiting Kapteyn. Kapteyn b may support liquid water on its surface. It completes its orbit within 48.6 days at an average orbital distance of 0.17 AU, has an orbital eccentricity of 0.21 and is around 11.5 billion years old. This makes it two and a half times older than Earth and just two billion years younger than the universe itself. Kapteyn c is more massive (seven Earth-masses) and its year lasts 121.5 days at an average orbital distance of 0.31 AU. These planets are in a dynamical state called an apsidal co-rotation, which implies that the planets are dynamically stable over long time-scales.

Gliese 832 is a red dwarf around 16 light years from Earth in the constellation of Grus. Its mass is half that of the Sun and hosts two known planets. In September 2008, astronomers discovered a Jupiter-like planet now designated as Gliese 832 b in a long-period, near-circular orbit. In 2014, astronomers at the University of New South Wales discovered a second planet, Gliese 832 c, which has an Earth Similarity Index of 0.81, one of the highest for any known extrasolar planet. Gliese 832 c orbits its star much more closely than Earth orbits the Sun. Since it orbits a red dwarf, it receives only as much energy from it as the Earth does from the Sun. Its temperature is predicted to be Earth-like, but with significant swings. The planet has a high eccentricity taking it very near to the predicted inner edge of the habitable zone.

This is what we know. Detecting more planets will become easier in the future. The European Southern Observatory's next-generation spectrometer ESPRESSO, which looks for Earth-like planets, will come

[24] M-class dwarf stars have attracted astronomers' attention. They are smaller than the Sun, are thrifty with their fuel and can last trillions of years, are more numerous than Sun-like stars, and are ideal in the search for habitable planets. However, they present a problem still being debated by astronomers. Their habitable zone is five times narrower than the one around the Sun. If a planet orbit is too close to a star, it can become tidally locked, thus presenting the same face inward. Such planets cannot be hospitable to life as we know it, but astrobiologists argue that liquid water may exist in the twilight zone between the face permanently exposed to its sun and the dark face.

[25] See Anglada-Escudé G. et al. (2014).

online in 2017 and will offer radial-velocity measurements several times more precise than those used up to now. Soon, both terrestrial and space-based telescopes will provide analyses of the spectrum of planets to identify chemical elements, such as oxygen, in their atmospheres. During the developmental phase of practical antimatter propulsion, astronomers will find many other planets, among them real targets for interstellar missions, that is planets in the habitable zone with Earth-like masses and oxygen in their atmospheres. These might be considered as 'our Americas' in space.

5.4 The Exploratory and Colonising Missions

Discovery of habitable planets within 5–15 light years from Earth will be the crucial driver for interstellar exploration and colonisation. It will provide the incentives to invest in technologies to produce antimatter at reasonable costs. However, before despatching human missions, robotic ones should be organised with the goal of gathering information that astronomy cannot give us, in particular the existence of magnetic fields, which shield planets from cosmic radiation, thus making them suitable for colonisation. Small robotic probes will land on planets to assess the soil and air composition, and other conditions, for the survival of small human settlements. Robotic missions will ascertain whether life forms, including intelligent life, exist on a planet. We believe that intelligent life does not exist on extrasolar planets near Earth, since the search for emitted signals have so far yielded no result.[26] But no one knows what lies around the corner. The first wave of colonists would face the challenge of surviving until the spaceship's return from Earth. They would need the same psychological and professional profiles as those selected for Mars colonization. Their number would depend on how many years the colony would remain isolated before the spaceship's return to Earth.

[26] The first attempts to search for intelligent life began more than half a century ago. Frank Drake started project Ozma and continued with the more ambitious and sophisticated project SETI (Search for Extraterrestrial Intelligence). SETI searches for intelligent life with radio telescopes to detect their signals. These projects analysed signals from many stars near the solar system, but there has only been absolute silence.

Researchers in conservation biology adopt a "50/500" rule of thumb. To prevent unacceptable rates of inbreeding, this rule requires a short-term population size of 50. The same rule needs a long-term population size of 500 to maintain overall genetic variability.[27] We thus assume that the number of colonists should be between 50 and 100. This would enable normal reproduction for around 30–60 generations.

What will settlers bring with them? It depends on the particular planet's characteristics as identified by robotic missions. Assuming that these characteristics are Earth-like, settlers will carry equipment to: (i) extract minerals necessary to build the colony; (ii) produce energy, which can be solar or nuclear; and (iii) produce fuels or other sources of energy for use by surface vehicles. They will bring enough seed stocks to produce agricultural commodities before the colony becomes autonomous. Lastly, they will bring 3D printers to manufacture equipment and other products necessary for the survival of the colony.

The mission's financing would follow the pattern used for the colonisation of Massachusetts.[28] Let's see how it works. A group of young persons willing to invest enough money will set up a Deep Space Company (DSC). Remember that the space economy has extended to Mars and the asteroid belt, and Earth is immensely richer than it is today. Individuals (the settlers), tired of living on the overcrowded Earth, look to move to the new found planet. They then petition the United Nations to govern the colony according to a statute governed by its laws. The DSC agrees with the settlers that, in exchange for their passage, the colony will grant DSC the rights of mineral exploitation on the new planet.

The DSC would finance the expedition for an amount equal to 55 % of the fuel and spaceship costs, say \$2.3 zillion. The settlers, through crowd financing, buy the equipment and more. The DSC then deposits

[27] A population size of 50 corresponds to an inbreeding rate of 1 % per generation. This is half the maximum rate tolerated by domestic animal breeders. The population size of 500 balances the rate of gain in genetic variation with the rate of loss due to genetic drift.

[28] The Pilgrim Fathers were a small group of English Puritans. Most of them were farmers, poorly educated and without social or political standing. In 1617, discouraged by economic difficulties and their inability to secure civil autonomy, the congregation voted to emigrate to America. Unable to finance the costs of the voyage, they organised a joint-stock venture. They negotiated a financial agreement with a group of London businessmen, who provided the capital. The ship began its historic voyage in 1620 with 102 passengers. See Labaree B. (1979).

$two zillion with an international public bank, such as the World Bank, for the mission's duration. The interest rate is 4 % per year. The mission lasts 30 years in the spaceship's reference frame. Taking into account time dilation, time passed on Earth is 70 years. Thus the end value of the deposit will be around $31 zillion. The DSC issues zero-coupon bonds, which are bought at a price lower than their nominal value, which is paid in full at expiration. The maturity of these zero-coupon bonds is 70 years. The DSC decides that the price of zero-coupon bonds is $1.8 zillion. Since these bonds are free of default risk due to the guaranteed bank deposit, we assume an interest of 4 %. The nominal value of the zero-coupon bonds is around $28 zillion.

The mission is now ready to go, with the young investors travelling with it. On arrival, the spaceship will remain in orbit until an outpost is established on the new planet. Upon return to Earth, the DSC pays off the zero-coupon bonds, realises $3 zillion surplus, and owns the spaceship. Since the time elapsed for the young investors is 30 years, return on investment is around 85 %, excluding the value of any cargo they may bring back.

Make a leap in time and several individuals in the new colony leave the planet for another one which is 5–15 light years distant. Over the centuries, humankind will have populated part of the galaxy. Such migrations resemble those of the Polynesians. Historians[29] of the Polynesian people, before European contact, estimate that their journey stretched from 3000 BC to around AD 800 (Fig. 5.3). Around 3000 BC, speakers of Austronesian languages, probably on Taiwan, mastered the art of long-distance canoe travel and spread south to the Philippines and Indonesia, and east to Micronesia and Melanesia. They then branched off and occupied Polynesia to the east. Dates and routes are uncertain. They started most likely from the Bismarck Archipelago and reached Samoa and Tonga around 1500 BC. By 100 AD they were in the Marquesas Islands and 300–800 AD in Tahiti and Easter Island, their easternmost point. The same date range is estimated for their arrival in Hawaii, far to the north and distant from these other islands.

---

[29] See Kirck P.V. (2000).

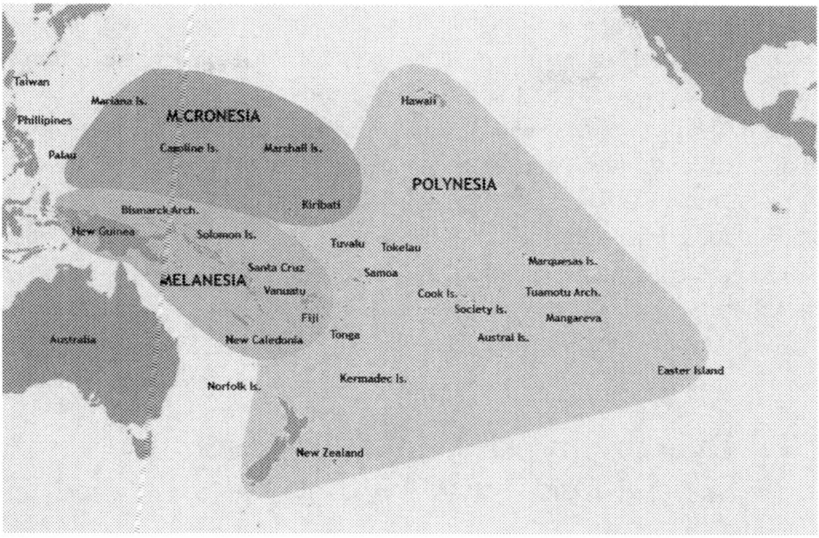

Fig. 5.3 Migrations of the Polynesian people (*Source*: https://commons.wikimedia.org/wiki/File:Pacific_Culture_Areas.jpg)

It is evident that these voyages were not made by chance. The transfer of many species of crops and livestock all over the Pacific islands prove that the settlements were made by colonists carrying products and animals from their homeland, deemed essential for the survival of the new colonies. There is still a debate on the motivations for these migrations. Several arguments have been put forward. One refers to population growth in the new settlement that may have put pressure on the limited agricultural resources of the island, inducing a group of colonists to move either into the unknown or to new islands previously discovered by voyagers sailing upwind on a predetermined bearing. Another argument is that human activities, such as deforestation in the new settlement, placed pressure on a fragile ecosystem thus creating a crisis that could only be solved by migration. In reality, it was probably a combination of these factors that was behind the spreading of the Polynesian people throughout the Pacific Ocean.

These people populated the Pacific Ocean over a period of around 4000 years. They jumped from one island to another, braving the unknown and the dangers of sailing the ocean with small boats.

Similarly, humankind will populate the galaxy over centuries and will inhabit semi-isolated islands in space, which will increase in diversity, both culturally and genetically. Astronauts will form a different civilisation and connect these islands by guaranteeing trade between them. Time dilation helps them to lose emotional and other personal ties with the inhabitants of various other islands in space. The special theory of relativity imposes these limits.

5.5 Overcoming the Limits Imposed by Special Relativity: Myth or Reality?

Limits imposed by the special theory of relativity are: (i) nothing travels at a speed greater than light; and (ii) clocks on board spaceships travelling at significant fractions of the speed of light slow down compared to the clocks of observers at rest. Is it possible to overcome these limits? An answer is given by the general theory of relativity, which allows space–time warping. This is no fantasy and can be verified by astronomical observations.[30] The general theory of relativity requires us to be more precise in assertions about motion. Instead of saying that nothing can exceed the speed of light, it states that spaceships travelling in space cannot exceed it. If space–time is warped, this limit no longer applies. If space–time is locally deformed, it shrinks before and expands behind spaceships, which are pushed forward by space–time as surfboards ride on waves. They never travel locally at speeds greater than light, since light is transported by the space–time expansion waves.

[30] The first test took place in 1919. Two British astronomical expeditions confirmed that the Sun's mass deforms space according to the general theory of relativity. Other scientists have confirmed that time is under gravity's influence. Clocks at high altitude slow down compared to clocks at sea level. Other more sophisticated tests have consistently confirmed space–time warping as predicted by the general theory of relativity.

Miguel Alcubierre[31] worked on a theoretical demonstration in which he proposed one space–time configuration in which a spaceship travels between two points in an arbitrarily short time. Other studies[32] have shown the theoretical possibility of distorting space–time geometry so that immense gravitational fields are not large near the spaceship and star bases. In this bubble, immense tidal forces due to gravity do not create complications. Space can be nearly flat and clocks on the spaceship and star bases stay synchronised. But there is a huge gap between saying "this violates none of the known laws of physics" and "we have detected a bubble of space–time curvature in the real world." NASA is financing experiments to measure the bubbles theorised by Alcubierre. Harold White made the first experiment to test Alcubierre-drive effects, which used an interferometer to measure the effects at the scale of nanometres.[33] So far the data are inconclusive. While experiments indicated an effect different from zero, external sources might have caused this difference. Other experiments are needed.

The good news ends here. The research mentioned above is related to the left-hand side of Einstein's equations, which, in a heuristic way, can be represented as follows:

$$\underset{(\text{Space-time warp})}{\text{Left members}} = \underset{(\text{mass-energy distribution})}{\text{Right members}}$$

The equations' left-hand members set the space–time geometry, while the equations' right-hand members set the mass-energy distribution. What sort of mass-energy distribution is required to obtain the space–time distortion needed for interstellar travel? To obtain this space–time configuration, Alcubierre's proposed mechanism implies a negative energy density. This is not surprising. To travel great distances in an arbitrarily

[31] Alcubierre A. (1994).

[32] See White H. G. et al. (2006).

[33] The original device proposed by White is an interferometer modified by using a laser beam. The beam is divided into two paths. The curvature of space causes a phase shift on each laser beam that should be detectable. Researchers have tried to learn if space deformation caused by the energy of an electric field with high voltage (up to 20 kV) could be detected. Plans are to increase sensitivity up to a hundredth of a wavelength and to create an oscillating field for more accurate results.

short time, the solutions of general relativity equations imply that matter must gravitationally repel other matter; and a theorem states that this condition is equivalent to the energy of matter being negative.

At present, we do not understand why negative energy exists or what its origin is. At the macro-level, negative energy exists. In 1998, observations of the rapid expansion of the universe[34] confirmed the existence of negative energy. This finding is counter-intuitive since gravity exerts an attraction and should decelerate any expansion. The acceleration can be explained on the assumption that 70 % of the universe's total energy is negative and 30 % is matter. At the microscopic level, quantum theory informs us that space is not empty. Quantum mechanics and the special theory of relativity imply that the distribution of energy in space is negative. Merging quantum mechanic's vacuum condensation with general relativity results in negative energy wiping out the universe a few moments after the Big Bang. The universe would then be greatly different. This puzzle remains one of the crucial problems of theoretical physics. It suggests our incomplete understanding of the vacuum and/or of gravity.

Many theoretical physicists are working on a quantum theory of gravity. If such a theory is found, it will answer the questions as to why negative energy exists and what its origin is. It will also offer ways to produce negative energy. Before Coulomb, Faraday, and Maxwell, scientists considered electricity and magnetism to be two separate phenomena. No one had the faintest idea how to produce electricity. After Maxwell's four equations merged electricity and magnetism together, we learned to produce electricity on an industrial scale. The second industrial revolution took place in the second half of the nineteenth century. The inventions of communication devices (e.g. telegraph, radio, and telephone) revolutionised our ways of living and doing business. These developments bear witness to the importance of understanding key physical phenomena systematically and theoretically.

[34] Astronomers used the stellar explosions or supernovas in distant galaxies to calculate their distances. When they measured the speed of recession of those galaxies, they discovered, to their surprise, that the universe's expansion is accelerating.

Even if someone elaborates a quantum theory of gravity, the energy needed to adapt space–time configurations for interstellar travel is extremely large. The equations of the general theory of relativity incorporate a constant that stipulates the elasticity of space–time. Einstein determined its value.[35] He found this coefficient equal to 2×10^{-48}. The inelasticity of space–time, given by the inverse of this ratio, is large. Hence, the energy required to distort space–time is large. The question is whether it is possible to achieve the huge energy needed for interstellar flight, which many physicists believe to be impossible using space–time distortion. We are sceptical, but keep our minds open. Eminent theorists[36] have argued on many occasions against phenomena proposed by general relativity. Later on, they reneged on their statements due to hard contrary evidence.

In 1964, the Russian astrophysicist Nicolai Kardashev[37] suggested that the technical progress of a civilisation is directly related to the energy its citizens can manipulate. According to this theory, culture (in the broadest sense) is a product of energy and technology: through technology, energy is used; since social and philosophical systems are expressions of technology, culture rests on and is determined by the energy used. Kardashev proposed three levels or types of civilisation:

- *Type I: Planetary Culture.* This uses all available resources of power in the world (in our case 10^{15} watts). Our civilisation is close to this type.
- *Type II: Stellar Culture.* This civilisation is much more advanced than ours and exploits the full power of its star (in our case, around 10^{26} watts).

[35] Einstein assumed that, with approximation, these equations restore Newtonian theory starting from the gravitational potential. He found that the coefficient of elasticity of space–time is equal to $8\pi G/c^4$, where G is Newton's constant and c is the speed of light.

[36] In the 1930s J.R. Oppenheimer presented a paper on black holes at a conference in Brussels. He hypothesised that objects known as neutron stars could not exceed two times the solar mass without collapsing into a black hole. An equally eminent theoretical physicist, J.A. Wheeler, argued that this result was impossible, since the laws of physics protect the heavenly bodies from such an absurd end. Ten years later, Wheeler had to renege on this statement, and, by curious irony, it was he who gave the name blacks holes to these objects.

[37] See Kardashev N. (1964).

- *Type III: Galactic Culture*. This civilisation harnesses the power of a galaxy, which is 100 billion times the energy produced by a civilisation of type II, more or less the same order-of-magnitude jump separating type I and II civilisations.

We believe type III civilisations will be able to travel between stars using a warp engine. If we learn to produce antimatter efficiently, humankind will expand into the galaxy over the next centuries. It is even reasonable to imagine that we will become a type III civilisation. At that point, humanity, dispersed in semi-isolated islands in space, could reunite into a global galactic society. Myth or reality? In more or less 140,000 years, *Homo sapiens* dispersed, travelling on foot, to the four corners of the Earth. During this dispersal, different civilisations and cultures emerged, isolated from one another. Millennia later, humanity reunited. Today we are living in a global society. Our own experience creates, so to speak, a precedent, or, paraphrasing the poetic expression of Mark Twain, “history does not repeat itself but it rhymes.” Seen in this perspective, the conquest of the stars appears in a different light, as a possibility. Remember the words of Einstein, “all that is possible is real.”

5.6 Appendix: Equations for Near-Relativistic Velocities

The specific impulse of relativistic rockets is the same as the effective exhaust velocity: despite the fact that the non-linear relationship of velocity and momentum as well as the conversion of matter to energy have to be taken into account, the two effects cancel each other:

$$I_{\mathrm{sp}} = v_e$$

This is only valid if the engine does not have an external energy source (e.g. a laser beam from a space station, in which case the momentum carried by the laser beam also has to be taken into account). If all the energy to accelerate the spaceship comes from an external source (and there is no additional momentum transfer), then the relationship between the effective exhaust velocity and the specific impulse is as follows:

$$I_{\text{sp}} = \frac{v_e}{\left(\text{sqr}\left(1-\left(\frac{v^2}{c^2}\right)\right)\right)} = \gamma_e \cdot v_e$$

where γ is the Lorentz factor.

In the case of no external energy source, the relationship between I_{sp} and the fraction of the fuel mass η which is converted into energy might also be of interest; assuming no losses,

$$\eta = 1 - \text{sqr}\left(1-\left(\frac{I_{\text{sp}}^{\;2}}{c^2}\right)\right) = 1-\left(\frac{1}{\gamma_{\text{sp}}}\right)$$

The inverse relation is

$$\text{I}_{\text{sp}} = c \cdot \text{sqr}\left(2\eta - \eta^2\right)$$

Here are some examples of fuels, the energy conversion factors, and the corresponding specific impulses (assuming no losses):

Fuel	Energy conversion factor	I_{sp}/c
Matter–antimatter annihilation	1	1
Nuclear fusion	0.00712	0.119
Nuclear fission	0.001	0.04

In actual spaceship engines, there will be losses, lowering the specific impulse. In electron–positron annihilation, the gamma rays are emitted in a spherically symmetric fashion, and they almost cannot be reflected with current technology. Therefore they cannot be directed towards the rear.

In order to make the calculations simpler, we assume that the acceleration is constant (in the spaceship's reference frame) during the acceleration phase; however, the result is nonetheless valid if the acceleration varies, as long as I_{sp} is constant. In the non-relativistic case, one knows from the (classical) Tsiolkovsky rocket equation that

$$\Delta v = I_{sp} \ln\left(\frac{m_0}{m_1}\right) \tag{5.4}$$

Assuming constant acceleration *a*, the time span *t* during which the acceleration takes place is:

$$t = \left(\frac{I_{sp}}{a}\right) \ln\left(\frac{m_0}{m_1}\right) \tag{5.5}$$

In the relativistic case, the equation still valid if solving this equation for the ratio of initial mass to final mass gives:

$$\left(\frac{m_0}{m_1}\right) = \exp\left(\frac{at}{I_{sp}}\right) \tag{5.6}$$

where exp denotes the exponential function. Another related equation gives the mass ratio in terms of the end velocity Δv relative to the rest frame (i.e. the frame of the rocket before the acceleration phase):

$$\left(\frac{m_0}{m_1}\right) = \left[\frac{\left(1 + \left(\frac{\Delta v}{c}\right)\right)}{\left(1 - \left(\frac{\Delta v}{c}\right)\right)}\right]^{(c/2\,Isp)} \tag{5.7}$$

For constant acceleration, $\Delta v / c = \tanh\left[at / c\right]$ (with *a* and *t* again measured on board the rocket), so substituting this equation into the previous one and using the identity $\tan\text{h}\, x = \left(e^{2x} - 1\right) / \left(e^{2x} + 1\right)$ returns the earlier Eq. (5.3).

By applying the Lorentz transformation to the acceleration, one can calculate the end velocity Δv as a function of the spaceship-frame acceleration and the rest-frame time t'; the result is:

$$\Delta v = \frac{at'}{\left(\text{sqr}\left(1 + \left[\frac{(at')^2}{c^2}\right]\right)\right)} \tag{5.8}$$

The time in the rest frame relates to the proper time through the following equation:

$$t' = \left(\frac{c}{a}\right) \sinh\left[\frac{at}{c}\right] \tag{5.9}$$

Substituting the proper time from the Tsiolkovsky equation and substituting the resulting rest-frame time in the expression for Δv, one arrives at the desired formula:

$$\Delta v = c \tan\text{h}\left[\left(\frac{I_{sp}}{c}\right) \ln\left(\frac{m_0}{m_1}\right)\right] \tag{5.10}$$

References

Alcubierre A. The warp drive: hyper-fast travel within general relativity. Class Quantum Gravit. 1994;11(5):73–7.

Anglada-Escudé G et al. Two planets around Kapteyn's star: a cold and a temperate super-Earth orbiting the nearest halo red dwarf. Mon Not R Astron Soc Lett. 2014;443(1):89–93.

Artie P et al. Evidence for a long period planet orbiting Ephsilon Eridani. Astrophys J. 2010;544(2):145–8.

Bussard RW. Galactic matter and interstellar flight. Astro Acta. 1960;6:179–94.

Forward R. Antiproton annihilation propulsion. J Propuls Power. 1985;1(5):370–4.

Forward R. Indistinguishable from magic. New York: Simon & Schuster; 1995. Amazon Books.

Frisbee R. How to build an antimatter rocket for interstellar missions. AIAA Paper 4676, 39th Conference, Huntsville, Alabama; 2003 July.

Kardashev N. The transmission of information by extraterrestrial civilizations. Sov Astron. 1964;8:217.

Kirck PV. On the road of the winds: an archaeological history of the Pacific Islands before European contact. Berkeley: University of California Press; 2000.

Labaree B. Colonial Massachusetts: a history. Millwood: KTO Press; 1979.

Leavitt HS. 1777 variables in the Magellanic Clouds. Ann Harvard Coll Obs. 1908;LX(IV):87–110.

Lewis RA et al. Antiproton-catalyzed microfission/fusion space propulsion systems for exploration of the outer Solar System and beyond, In: El-Genk M, editor. Space technology and applications international forum; 1997 Jan 26–30; Albuquerque, New Mexico, CONF-970115, American Institute of Physics, New York; 1997.

Mayor M et al. A Jupiter-mass companion to a solar-type star. Nature. 1995;378(6555):355–9.

Santerne A et al. SOPHIE velocimetry of Kepler transit candidates. Astron Astrophys. 2012;575:76.

Schmidt GR. et al. Antimatter production for near-term propulsion applications. NASA Marshall Space Flight Center, Document ID:19990100915; 1999.

Tsiolkovsky K. Exploration of the universe with reaction machines: exploring the unknown. The NASA History Series, NASA SP 4407, Washington DC; 1995.

Van de Kamp P. Alternate dynamical analysis of Barnard's star. Astron J. 1969;74(8):757.

White HG et al. The Alcubierre warp space drive in higher dimensional space time. In Proceedings of Space Technology and Applications Forum, University of New Mexico and American Institute of Physics; 2006.

6

The Uncertainties

This book has developed a scenario for space exploration and colonisation that is hypothetical but realistic. Exploration of asteroids by space agencies will have been accomplished during the next ten years. The technical capabilities to do so already exist. Financing asteroid exploration will not require major increases in the budgets of space agencies above their actual and foreseen funding, if synergies with private companies are found. Asteroid exploration will pave the way for space mining by private companies, which could very likely occur in the late 2020s. Financing space mining could be achieved through the financial and equity markets, provided an international legal framework for the use of space resources is in place.

We even have the knowledge and technologies to explore and colonise Mars. Our scenario posits Mars exploration in the early 2040s. If space mining does develop, we estimate a 5–10 % reduction in global defence spending over the next 20 years. This will make financing Mars exploration possible and the planet's eventual colonisation. Most importantly, Mars exploration needs extensive international collaboration among space agencies and industries. To go beyond Mars requires the development of technologies that are not yet mature, while interstellar

G. Bignami, A. Sommariva, *The Future of Human Space Exploration*,
DOI 10.1057/978-1-137-52658-8_6

exploration requires advances in science and technologies that we do not yet have.

This chapter analyses the uncertainties surrounding our scenario that may push the future in different directions. It focuses on risk analysis of solar-system exploration and colonisation since interstellar exploration is too remote a possibility to make realistic risk analysis sensible. Human exploration and colonisation of the solar system needs international cooperation. The key question is whether our present civilisation is ready to embrace a culture of non-violence in managing international relationships, to reduce gradually military spending, and to devote resources to space exploration without disrupting other major governmental programmes. Hence uncertainties exist both in the political and cultural realms.

6.1 The Present State of International Relationships

In the early 1990s, the world became unipolar. The United States emerged as the dominant military, economic, and technological power. During Bill Clinton's presidency, starting in 1992, the United States' primacy in world affairs went unchallenged. Clinton devoted much of his foreign policy to helping Russia transition to a market economy and establish democratic institutions, and to solving conflicts in the former Yugoslavia, East Timor, Northern Ireland, and the Middle East, where the Israeli-Palestinian conflict dominated.

Clinton's economic policy focus can be encapsulated by the following four points: establish fiscal discipline and eliminate budget deficits; keep interest rates low and encourage private-sector investment; cut protectionist tariffs and promote free trade; and invest in human capital through education and research. His economic approach entailed modernising the federal government, making it more enterprise-friendly while dispensing greater authority to state and local governments. But his policies were not a total departure from the previous two Republican administrations' policies. He introduced and signed several deregulatory laws, among which, most notably in 1999, was the Financial Services Modernization Act. This act allowed banks, insurance companies, and investment houses to merge as it repealed the Glass–Steagall Act which had been in place since 1932. During the 1990s, moreover, these policies became widely accepted by major political parties in core European states.

During the following two decades, events in five regions changed the global distribution of power:

6.1.1 The United States

After the attack on the Twin Towers of Manhattan's World Trade Center, the Bush administration shifted its foreign-policy orientation to national security. It put emphasis on: the United States as a moral nation with a duty to act as a benevolent global hegemon; spreading democracy around the globe to create peace; distrusting international institutions and being sceptical of their effectiveness, thus leading to more unilateral actions by the United States. In reaction to this attack, the United States led a NATO invasion of Afghanistan, instigating the "Global War on Terror." NATO forces scoured the region for Osama bin Laden and his terrorist network al-Qaeda and drove the fundamentalist Islamic Taliban regime from power. Bush called for full implementation of the Iraq Liberation Act signed into law by Clinton in 1998. He claimed that the Iraqi government had ties to terrorist groups, was developing weapons of mass destruction, and did not cooperate sufficiently with United Nations' weapons inspectors. In early 2003, the war to oust Saddam Hussein was launched.

Global military involvement led to strong increases in military spending in the USA. At the same time, the Bush administration succumbed to the pressures of the economic and financial elites to reduce taxes. Increases in government expenditures and reduced revenues led to massive government deficits and mounting federal government debt. For a while, the increased absorption of the United States' government debt by emerging countries managed to stabilise these imbalances, since the emerging countries maintained a stable exchange rate of their currencies to the dollar. The stability of the exchange rates helped to maintain the competitiveness of their products in the US markets, as loose American monetary policies sustained private consumption and economic growth. But this led to rising twin deficits in the United States, both internal and external, and to several bubbles in the financial markets. The subsequent collapse of the real estate bubble led to the dramatic slowdown of economic growth and threatened to become a world economic slump.

At the start of 2009, the new Obama administration inherited two crises from the Bush administration: a domestic economic crisis that threatened to become a world-wide economic slump and mounting evidence of the failure of the recently established democratic Iraqi state. The administration gave priority to the domestic crisis. Expansionary monetary policies, devaluation of the dollar, and federal intervention to save the failing automobile industry succeeded in restoring economic growth. But economic growth is fragile. The United States cannot be the only locomotive of the world economy. Stagnation in the European Union and the financial crisis following the possible Greek exit from the eurozone threatens economic growth worldwide. Moreover, in the last decade, technical innovation in the United States did not translate into increased labour productivity, contrary to what happened in the 1960s. The average productivity growth in the non-farm business sector was around 3 % in the 1950s and 1960s. Between 2007 and 2014, productivity grew by only 1.4 %, despite the technical innovations coming out of Silicon Valley. If labour productivity does not increase, neither does the salaries of workers. The income gap remains high. A report by the Organisation for Economic Co-operation and Development (OECD) concluded that income inequality in the United States had reached its highest level for the past half century.

The new administration changed its foreign-policy strategy gradually. Important geopolitical events occurred, including: the aftermath of the 2008 world-wide "Great Recession" and the ensuing eurozone crisis; renewed dialogue with Iran; widespread Arab-Spring protests that toppled many governments in the Middle East and North Africa, precipitating civil wars in Libya and Syria; tensions with Russia pushing back against NATO and the European Union's expansion into areas once under control of the Russian and Soviet Empires; and rising tensions with China. As of 2016, China is challenging the United States through: its aggressive policies in Africa; territorial expansion in the South China Sea; establishment of an international fund competing with the World Bank; increased military spending; and a shift to strategic cooperation with Russia.

Obama's foreign policy can be described as based on: cooperation with allies; multilateralism; the end of the Iraq War; the destruction of the core leadership of terrorist groups; stopping the spread of weapons of mass destruction; and building more secure societies, not by direct intervention but by working with citizens of those societies. Several factors hampered the administration's foreign policy. Russian and Chinese aggressive policies raised the spectre of a new Cold War. So far, containment through economic sanctions has been the administration's response to Russia's policies. But economic sanctions work only if significant consensus with other nations exists for their implementation. Obama made America's pivot to Asia a centrepiece of his foreign policy architecture. But beset by crises elsewhere, the US's Asia pivot remains more rhetoric than reality. After promises of stronger US military presence in the Philippines, Singapore, and Australia, little evidence exists on the ground.

In summary, Obama's foreign-policy strategy is well articulated, but the problems are in the implementation. Moreover, partisanship in every aspect of American politics makes foreign policy implementation difficult. This affects the way other countries look at its foreign policy, raising questions about the administration's ability to implement a coherent strategy.

6.1.2 China

In the 1990s, production of many manufactured goods shifted to Asia. This has been a positive development since millions of people were lifted out of poverty as the region industrialised. Many Asian countries adopted policies tailored to develop the foreign-oriented economy, generating foreign exchange through exporting products and importing advanced technologies. Wages are kept low by state repression of labour organisations and restrictions of democratic political activities. These developments have attracted the strong interest of transnational corporations from the automotive to the electronic industries. They were looking for new bases where there is an abundance of skilled and cheap labour to diversify their global production network.

During this period, China, a major regional power, became a full member of the international community, a member of the United Nations Security Council and an active participant in the World Trade Organization. Despite some sabre-rattling rhetoric directed at Taiwan, China pursued an international policy leading to seats at the tables where the international rules are written. It did not manifest the aggressive foreign policies reminiscent of the Japan of the 1930s, and it did not challenge the global power of the United States. This was consistent with its overall strategy of political stability and prosperity.

In 2008, Chinese policy makers responded strongly to the global economic crisis. They launched an economic stimulus programme to offset the negative effects of the downturn in international trade. As China's growth rates slowed with respect to the recent past, several economic imbalances emerged: the continuous reliance on investment and exports to generate economic growth; the decline in households' disposable income as a percentage of GDP; the relatively low level of household consumption in GDP; and the presence of an outsized manufacturing sector. Over the past decade, negative real-interest rates, the emergence of a significant informal credit market, and the sharp decline of government bonds held by households are all indications of a repressed financial market. Liberalisation of interest rates and the exchange rate are central to the reform of the financial system and the reduction of the saving–investment imbalance.

In 2015, China has devalued the yuan and expanded public works. But there are some doubts as to whether this devaluation and expanding public works will jolt the economy back into action. This is due to what is happening in China's labour market (Fig. 6.1).

For many years, China's labour market has been tight as people moved from rural areas to cities in search of work. Beginning in 2010, many people have been aged out of the labour market, pushing the ratio of job offers to seekers upwards. This indicates that the Chinese economy has reached the limit of its labour-supply capacity, and economic stimuli, such as devaluation and public works, may have limited impact on

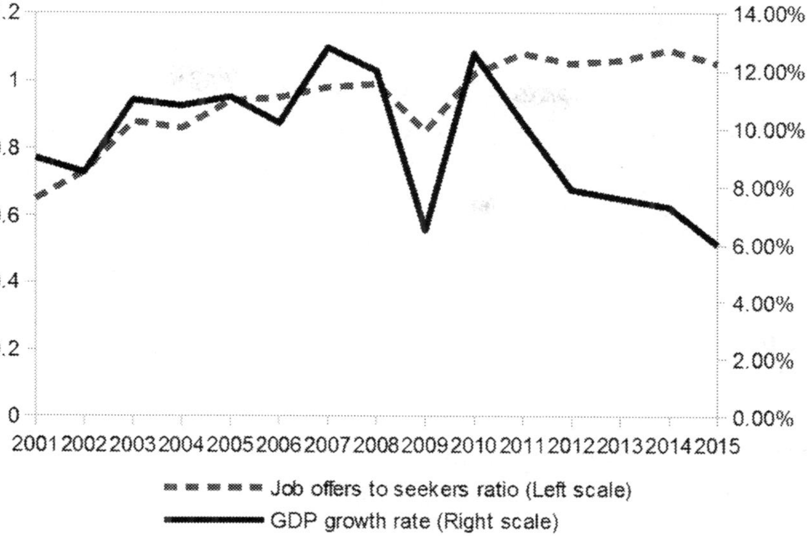

Fig. 6.1 China's labour market and GDP growth

economic growth. The only way to revive economic growth is through increased productivity per worker, which requires a more efficient allocation of resources. But it will take some years to make workers more productive and, in China's case, to liberalise the economy to achieve a more efficient allocation of resources.

Combined with the slowdown of international trade, economic imbalances can pose a threat to prosperity. Perhaps a yet more critical factor, the regime's legitimacy in mainland China, rests on prosperity. China's population, geographical vastness, and social diversity frustrate attempts to rule from Beijing. Central government leaders must increasingly build consensus for new policies among party members, local and regional leaders, influential non-party members, and the population at large. If prosperity is the glue that holds the regime in place, threats to it endanger political elites and reinforce those conservative elements within the government who have a more aggressive nationalistic view. This is manifested in a more aggressive foreign policy that openly challenges the United States.

6.1.3 European Union

Twenty-eight European countries adopted a common currency in fthe 1990s. At that time, few economists were concerned that, without a common economic policy, monetary union among divergent economies increased the risk of greater susceptibility to external shocks. But many policy makers in Europe considered these risks to be slim. Then a profound external shock occurred in 2008 after the housing bubble's rupture in the United States. To avoid collapse in the eurozone, Germany, the major power within the European Union, adopted the agenda[1] started by Margaret Thatcher in the 1970s. While we do not question here the validity or accuracy of the neoliberal assessment of the crisis of the 1970s, we do question whether that agenda is relevant today for solving the economic and financial problems facing European countries.

Political elites in the eurozone have delegated managerial functions to the European Union and to privatised "public" power (international capital markets). But this model does not cut bureaucracies' power: it privatises it. It insulates bureaucratic power from any form of popular sovereignty. People, regardless of nationality or political identity, are subject to discipline imposed by international organisations and global financial markets. They have no recourse to the political process (i.e. electoral) to hold these entities accountable. One result is disaffection, because electoral systems do not appear effective to further and protect people's interests. Governments present themselves to the people as having no choice but to sacrifice to a global reality. States have less ability to access their resources due to unregulated international capital markets. This threatens the welfare state and diminishes governments' legitimacy in its citizens' eyes. Radical right and left parties, who share anti-European Union persuasions, are gaining votes.

The austerity policies adopted by the European Union have resulted in stagnation and high unemployment. The expansionary monetary policy adopted by the European Central Bank has halted the declining trend of

---

[1] This agenda attacked the basic presumptions of the Fordist class compromise and the welfare state. Thatcher argued that these were too expensive, too collectivist, and too paternalistic. She believed that the welfare state was responsible for the stagnation, unemployment, and inflation that then plagued Britain and much of the world economy.

the eurozone economies. Recently, the slowdown of China's economic growth has resulted in declining demand for raw materials and imported products. In turn, this has affected economic growth worldwide. This is particularly troublesome for the European Union, which is the largest trading partner of China. In 2011, total trade accounted for $567 billion, with exports totalling $122 billion. According to the European Union trade monitors, Germany contributes one-third of the total trade volume between China and Europe. But, it is likely that the slowdown in China's growth rate will affect German exports less than those of other eurozone countries. The modernisation of China requires German products that demonstrate their comparative advantage in machinery, transport, equipment, and other manufactured goods. What's more, China particularly favours Germany's exports, as they have won a global reputation for quality.

The slowdown of China's growth will increase the economic divergence among countries in the eurozone. The obvious solution is a common economic policy and a further democratisation of European institutions. The issues of political and fiscal integration have been extensively analysed and are well outlined in the report signed by five presidents (Juncker, Tusk, Draghi, Schultz, and Dijsselbloem). This report could be a road map for the further integration of eurozone countries, and the report is waiting for a political sponsor to become operational. But high unemployment, the lack of a European identity, and problems raised by large migrations from Africa and the Middle East are threatening political integration. It is too early to conclude whether thisthat integration in Europe has come to a halt, but the risk is real. If Europe fails, it will become irrelevant as a player on the world stage.

6.1.4 Russia

Twenty-five years after the Soviet Union's demise, Russia is reverting to imperialistic policies. When Boris Yeltsin came to power in 1991, an optimistic wave hailing a new era of democracy and economic freedom pervaded Russia. Instead, everything went wrong. Its economy sank into deep depression by the mid-1990s. The 1998 financial crash further battered the economy. According to Russian government statistics, economic decline was far more severe than the Great Depression of the 1930s in the United States. Public health indicators show a dramatic corresponding decline. By 1999, the total population fell by three-quarters of a million people. Life expectancy for men dropped from 64 years in 1990 to 57 by 1994. Women's life expectancy dropped from 74 to 71. Both health factors and a sharp increase in young peoples' deaths from unnatural causes contributed to this trend.

One decade after the Soviet Union's collapse, Russians were unhappy with their country's direction. Enthusiasm for democracy and capitalism waned as a widespread perception spread that political and business elites enjoyed the spoils while average citizens were left behind. Opinion polls showed that around half of Russians (48 %) believed it to be natural for their country to have an empire while only 33 % disagreed. In contrast, in 1991, during the final months of the USSR, fewer (37 %) expressed a preference for an empire, while 44 % disagreed.

President Putin exploited these sentiments. His accession to office came at an ideal time, after the ruble's devaluation in 1998, which boosted the demand for domestic goods as world oil prices were rising. Since the Russian economy is heavily dependent on oil and natural gas exports, higher oil prices boosted economic growth. Putin confronted several influential oligarchs who had acquired large stakes of the enterprises that were previously state assets and allegedly through illegal schemes. Confrontations led to his regime's establishing control over Russian media outlets and oil-and-gas assets briefly owned by the oligarchs. Repressive measures against internal dissent have become common. On repeated occasions, Putin has stated that the Soviet Union's fall led to too few gains and too many problems for most Russian citizens.

In 2008, Russian relations with the West deteriorated; that year saw the war in South Ossetia against Georgia following the latter's attempt to take over that breakaway region. Russian troops entered South Ossetia and forced Georgian troops back, establishing their control of the territory. In autumn 2008, Russia unilaterally recognised South Ossetia's and Abkhazia's independence. In 2014, President Putin requested and received authorisation from the Russian Parliament to deploy troops in Ukraine in response to the local political crisis; they gained complete control over the

Crimean Peninsula within a day. Putin sought to strengthen ties with the People's Republic of China by signing the "Treaty of Good-Neighborliness and Friendly Cooperation" and building the Trans-Siberian oil pipeline geared toward growing Chinese energy needs. Russia is threatening to cut natural gas supplies to European countries to moderate their attitude toward the imposition of economic sanctions. Moreover, the country is supporting right-wing anti-EU political parties to try to weaken the Union from the inside.

6.1.5 The Developing World

The widespread Arab-Spring protests toppled many governments in the Middle East and precipitated civil wars in Libya and Syria that have since become failed states. Uneven economic development along ethnic or religious group lines, severe economic decline, and widespread corruption and criminality are common factors behind the fragility of many African countries. To measure state failure, the Fragile States Index (FSI) is particularly important and has received comparatively great attention since its first publication in 2005. Edited by the magazine *Foreign Policy*, the ranking examines 178 countries based on analytical research of the Conflict Assessment System Tool (CAST) of the Fund for Peace. The top 20 countries in the 2014 FSI (i.e. those with the highest state-failure measures) include many African and Middle Eastern countries.

Failing states are breeding grounds for terrorism and can cause large population movements. From Libya to Syria and Iraq, failing states provide the opportunity to radical religious movements and terrorist groups to reassert their presence. They act as magnets to disfranchised locals and young Muslims from Europe and other developed countries. What's more, pressures from environmental, sectarian, and economic developments in weak states are causing large population movements. Today we are witnessing at least several mass migrations: from Bangladesh and Myanmar to Thailand and Malaysia; of Africans and Arabs to Europe; and of Central Americans to the United States. The United Nations Agency for Refugees reported that in 2014 population movement has involved more than 50 million people.

To check the negative impact of countries where disorder reigned, we relied in the past on empires, colonisers, and dictators. But we live in a post-colonial, post-imperialist, and post-dictatorial age. An obvious solution is to reduce the gap between rich and poor nations, which remains wide.[2] But, in our global economy, three serious issues undermine this solution. First is the high level of uncertainty brought about by financial markets, which is a consequence of speculative markets being divorced from any real economic processes. Second, increases in efficiency say nothing about the redistribution of wealth. Within one state, isolated from the world economy, redistribution can take place, but, in our globalised world, no mechanism exists for ensuring broader redistribution. There is nothing in the free-market mechanism which automatically encourages redistribution world-wide. Third, economic actors have great power, and their power over states is increasing. Lord Acton's aphorism, "power corrupts, but absolute power corrupts absolutely," applies to economic actors as well as to anyone else. One weakness of the economic liberal position is its neglect of power in the analysis of economic actors.

In summary, the rapid progress of globalisation and free trade have created conditions for economic growth world-wide. But in the United States, the agenda of deregulation paved the way to financial crisis and the resultant slowdown of the world economy. Despite the recent progress of the United States economy, world recovery is still fragile. Lack of global governance undermines efforts to manage weak states, and perceived terrorism threats have strengthened key aspects of state power in Western countries. Neo-conservatives have reinvigorated the debate over the separation of the private and public realms by representing the family as private where individual responsibility, patriotism, and traditional "family" values can be nurtured. As Arnold J. Toynbee once remarked, "nationalism is a sour ferment of the new wine of democracy in the old bottle of tribalism." Perceiving a weakening of the United States' global influence, Russia and China aggressively pursue their national agendas. Nationalism is surging worldwide.

---

[2] According to estimates of the United Nations, 66 % of world GDP is concentrated in Europe and North America, with Asia at 24 %. Africa scores badly with 1 % of world GDP.

6.2 The International Environment and Human Space Exploration

This book has set forth the case for the dependence of human space exploration on international cooperation. But the present international environment looks unfavourable to space cooperation. China has announced its intention of building its own space station. Russia will continue to cooperate on the ISS until 2024, but has announced its intention of going it alone after that date. Military planners, particularly in the United States and China, are considering how best to meet their future national security needs in space. Future needs include the safety of one's own citizens and assets in space, protection against piracy, and national security issues. Under the current international climate, the risk of the militarisation of space is real. While the risk of an immediate military presence in space remains low, it becomes higher in the next 10–20 years.

If competition for power prevails, our space scenario will not materialise. Major spacefaring nations may embark on manned missions to Mars. The United States and China have the technology and the resources to launch such missions. Recent studies by NASA estimate the costs at between $100 and 200 billion. This budget is affordable for the space agencies of the United States and China, pursuing the primary goals of finding life on Mars and demonstrating and growing the superiority of their technologies. But these missions risk the same fate of the Apollo programme: public spectaculars with few long-term benefits and sudden termination after achieving their goals. In summary, they will be a colossal waste of money.

The presidential election in the United States and general elections in many European countries will take place during the next two years. The international economy and the management of relationships will likely be among the foci of debate. International cooperation in space should be an important issue and three items in particular should be addressed: international legislation on the use of space resources, space mining, and initiating a cooperative exploratory mission to Mars.

6.2.1 International Legislation on Use of Space Resources

In the short to medium term, international collaboration is needed to introduce an international legal framework for the use of space resources. Will the reassertion of nationalism affect negatively international collaboration on the legal framework? We do not think so. In the present international climate, major spacefaring nations could be tempted to enact their own legislation on space mining. But such legislation must abide with the United Nations Outer Space Treaty signed by the major spacefaring nations. It is thus in the interest of these nations to work with the United Nations to revise the Outer Space Treaty and adopt national regulations for their own space-mining companies.

The enactment of the US Commercial Space Launch Competitiveness Act could be viewed as an opportunity. If the very discussion of asteroid mining has the power to influence, say, Russia's behaviour in the Arctic for the worse, could not it also be used to influence it for the better? And what about other countries with whom the United States has had recent communication breakdowns? In fact, since day one, space has always been about foreign policy. President Kennedy used the Apollo programme to demonstrate the superiority of US technology and to influence non-aligned nations. The Apollo–Soyuz Project of the 1970s was a demonstration of détente. In the early 1990s, the Clinton administration used international cooperation, especially with Russia, on the ISS to keep the nuclear weapons of the former Soviet Union from falling into the wrong hands. In the chaos following the collapse of the USSR, the US administration brought the Russians into the space-station project because the people that were in charge of the nuclear material were also in charge of the rocketry. They were the same people that would have to make the decision on the space station. So the strategy was to open a bona fide line of communication with the people in control of the bombs. It was a gamble. None of the other ISS partners initially wanted the Russians involved, but it worked.

The enactment of the US Commercial Space Launch Competitiveness Act could present new opportunities, if the United States engages in a dialogue with other space actors and the international community on issues concerning particular shortcomings in the Act or holes in the international legal framework. A new space race between the United States, China, Russia, and possibly Europe could then be kicked off with agreed-upon codes of conduct as well as the driving force of competition. Such activities might also direct focus away from the Arctic and the South China Sea to near-Earth asteroids and the Moon. So, the political opportunities that could come from discussions with other nations about space mining might be of immediate value. Political alliances forged over a discussion of space mining could yield benefits today and in the future in terms of a more stable geopolitical environment.

6.2.2 Space Mining

The knowledge economy is an important part of our economies. It is based on technological innovation, which is in turn generated by scientific research. In "Science, The Endless Frontier," a July 1945 report[3] to the president, Vannevar Bush maintained that basic research is "the pacemaker of technological progress." Bush's arguments can be summarised in three points: the new economic frontier is based on technological innovation; basic science is the mover of innovation; and governments must finance basic science to feed the whole innovation process. Bush's recipe is more timely than ever. But it has a missing part. There is no efficient linear passage of technological innovation from laboratories to industries. Therefore, governments should not only invest in scientific research, but they must create demand for the resultant high technology.[4] Innovations

[3] Bush V.: "*Science, the Endless Frontier: a Report to the President.*" Washington, D. C.: U.S. Government Printing Office, OCLC 1594001.

[4] In the 1950s and 1960s, this demand came from space exploration, which was instrumental in the birth of microelectronics and the space economy. The Apollo programme generated over half of the demand for integrated circuits, which gave a strong impulse to the nascent microelectronic industry. The Soviet Union's success with the Sputnik spurred the US Defense Department to launch a research and development programme for long-distance satellite-based communications. As a result, public spending paid for research and development. Moreover, launches of military satellites reduced the business risk for future private activities. In 1968, the companies taking part in the Defence Department programme launched the first civilian long-distance telecommunications satellite. Afterwards, an increasing number of countries entered the market. In 2011, the turnover of the satellite industry amounted to $177 billion. The growth of the industry's turnover remained

occur at the frontiers' outer edges, facing unprecedented challenges both cultural and technological. Today, space mining, Earth knowledge, and sustainable management of its resources are these frontiers' edges.

The United States is well advanced in a new private–public relationship that makes space mining possible. But Europe is lagging behind and has remained tied to the old model, characterised by state procurement contracts and a monopolistic launching industry. With the exception of Richard Branson, who is also interested in the development of space tourism, no private investor has expressed interest in developing space-mining activities. At the moment, the only company in Europe producing medium to heavy rockets (Series Ariane) is Arianespace. These are for the launch of commercial satellites. It has a close relationship with ESA for the launch of its satellites and its scientific missions. But competition from the United States, Russia, and China risks marginalising the old continent in these developing opportunities.

What can and should Europe do? We start with what should be done. First, a clear commitment at the European level for "planet hopping" is necessary. This would involve some radical institutional changes. At the moment, ESA is an institution coordinating the space programmes of participating European countries. This must be transformed into a federal institution along the lines of the European Central Bank. It should have the mandate to develop a "planet hopping" programme in coordination with the space agencies of other main spacefaring nations. This is not only a technical challenge but involves a debate at political and cultural levels over whether we want to advance towards a spacefaring civilisation or not.

As for the first hop, ESA and the European Commission should consider the deregulation taking place in the United States. They should design new policies to increase competition in the launching sector as a decisive step towards the development of low-cost vehicles for orbital flights and the independence of the European space sector. An exploratory programme of the asteroids should be carried out. ESA has already begun programmes for the exploration of asteroids and comets, but they

---

strong despite the adverse economic environment. In 2013 the turnover reached $239 billion, which is an increase of 35 % over two years.

have mainly scientific purposes such as plotting the orbits of the asteroids on possible collision courses with the Earth. An extra dimension should be added to these programs, i.e. the identification of asteroids suitable for future mining operations. Research and development on nuclear propulsion need also to be carried out, without which asteroid mining will be difficult to realise. Finally, the European institutions, in cooperation with other spacefaring countries, should work at the United Nations level to revise international legislation on property rights and uses of space resources.

The above outlined policies will have important effects on investment and economic growth in the eurozone. The money invested in asteroid exploration and in developing nuclear propulsion will be spent here on Earth generating employment and economic growth. The announcement of the intention to proceed with the exploration of asteroids in partnership with private space companies, and to collaborate with other nations in the establishment of an international legal framework for the use and extraction of space mineral resources, would send important messages to private investors and financial markets in Europe, creating the right incentives for private capital to enter into space activities. The most important message is that space activities will return economic, technological, and quality-of-life benefits on Earth. Finally, it would allow Europe to sit at the table of negotiations with other spacefaring nations with a position of strength. If Europe does not develop a coherent vision of space policies and implement them, it risks being marginalised in a sector that will be crucial for innovation and economic growth in the near future.

6.2.3 Mars Exploratory Mission

The Mars exploratory mission can be realised by a consortium of space agencies under the United Nations' aegis. International cooperation offers more efficient ways to achieve this mission, with the potential to reduce costs by spreading them across several nations. Although the costs of international cooperation increases the overall cost, the latter is divided amongst the partners. As costs for each partner decrease, so the utility for each partner increases. Benefits from cost efficiencies and sustainability will go to partners who approach space exploration as an effort of mutual benefit. International cooperation brings with it diplomatic prestige. More participating nations increase the diplomatic influence of each nation taking part. As a result, the larger the number of countries taking part in the programme, the greater the utility derived from cooperation. Once cooperation is established, cancelling the programme becomes incompatible with political sustainability. Costs associated with the loss of benefits caused by ending an international agreement are greater than those of continuing to take part in it.

We suggest three arguments for debate. *First*, a space-exploration programme, as outlined in Chaps. 3 and 4, can contribute to reviving productive investments, productivity, employment, and growth. Not only will the jobs and revenue engendered by a Mars exploration be of an order of magnitude higher than that brought by the Apollo programme,[5] but the economic effects of a cooperative Mars exploratory mission would be world-wide due to industrial cooperation among several countries and industries. Building successful cooperation among industries can increase productivity. It helps managers to track labour factors in the original scope of work to reflect conditions used to estimate and fund the project and to eliminate or decrease impacts on productivity that directly affect costs. Moreover, interplanetary exploration technologies will have spill-over effects in a variety of fields. These will include new materials, 3D printing, nanotechnology, computer technologies, communication technologies, biology, and also positive effects on productivity in other industries. In summary, a human exploration programme would be a powerful tool in the hands of governments to strengthen economic growth world-wide.

Innovation in nuclear propulsion, including nuclear-fusion, needed for human space exploration will have important economic effects on Earth and would enhance the reduction of liquid fuels and gas currently

[5] The Midwest Research Institute study of the relations between R&D expenditure and technology-induced increases in GNP indicated that each dollar spent on R&D returns on average over $7 in GNP, over the 18-year period following expenditure. Assuming the Apollo programme's R&D expenditure had the same economic pay-off as average R&D expenditure, $25 billion (1958) spent on Apollo R&D during 1959–1969 returned $52 billion through 1970. These research expenditures then continued to stimulate the economy through 1987, to a total of $181 billion.

consumed in the production of electricity. New energy sources such as nuclear fusion and space-based solar energy can address the complementary needs for the reduction of greenhouse-gas emissions. The physical impacts of climate change are complex and difficult to predict, but will certainly include higher global average temperatures, the melting of ice caps, rising sea levels, and the increasing scarcities of agricultural land and freshwater. Regions where soil is fragile, because of intensive farming, water logging, and wind erosion, are more sensitive to climate change, while the soils in at least several developing countries already display marginal physical characteristics. Further, tenant farmers in these countries have little incentive to care for the land as property rights are poorly defined. If climate change adversely affects agriculture, the human effects are likely to be more severe in the poorer world where more people are already at or near hunger. The Intergovernmental Panel for Climate Change[6] estimates that, to keep the world-temperature increase below 2 °C, at least 80 % of the world's electricity production must be low carbon by 2050. This is a massive global challenge that requires the exploitation of every available low-carbon energy technology.

In summary, space exploration programmes positively affect worldwide economic growth, productivity, and employment. Space technologies contribute to the mitigation of climate change and its effects on the development of poor countries. Together, space exploration programmes and associated improved technologies could result in more equitable and sustainable economic growth.

Second, sustained economic growth may help bring China and Russia back to the negotiation table. Recent opinion polls show widespread support in Russia for specific features of democracy, such as a fair judiciary, honest elections, freedom of the press, freedom of religion, free speech, and civilian control of the military. But the support for democracy is overwhelmed by fears of decline in living standards, law and order, and public morality. A new model of economic growth based on industrialisation, rather than the export of raw materials, may stimulate support for international cooperation by the more progressive elements in

[6] See IPCC Fifth Report on Climate Change 2014. This Report is available online at the IPCC official website.

Russian society. In China, a robust rate of world-wide economic growth may bring back to power elites favouring a seat at the table where international rules are written.

Nuclear-fusion and space-based solar power increase energy security, which, along with diversifying the sources of strategic minerals, space technologies, and mining, could contribute to global political stability. Although quarrels between nations will continue to exist, two of the above-mentioned powerful sources of controversy may be mitigated. The United States, as the leading military, technological, and economic power, is in a position to exert strong influence. Its willingness to collaborate with other nations to establish an international legal framework for the use of space mineral resources and to implement a cooperative Mars exploration programme would send critical messages to policy makers, emerging private space companies and financial markets, and other spacefaring nations. Such a collaborative policy could help progressive elites and populations in Russia and China to understand that peace and plenty have more to offer than confrontational policies and, accordingly, to re-evaluate their present aggressive policies.

Third, new frontiers can help humanity exit from irrationality and intellectual torpor. The discovery of the Americas revived instincts for exploration of a population then confined to narrow Europe. Testimonials about the distant lands and maps distributed through the new press fuelled humanism and curiosity. They ushered in a new era of scientific and intellectual research. The newly discovered continent's wealth stimulated entrepreneurs to take risks. It created conditions for the revival of the economies of European nations. The same rationale for exploring and settling space mirrors the spirit that has compelled explorers throughout the ages, that is the human urge to expand the space where they live and work and the frontiers of knowledge. That is the basic reason people explore and has been so since humans first walked on Earth.

By widening the horizons and future targets for humanity's exploratory instincts, science and technology prevent human civilisation from falling into apathy and stagnation. A new frontier would rekindle ordinary people's interest in space. This general interest spans a broad set of ideas and motivations, ranging from exploration as a human imperative to the possible evolution of human consciousness concomitant with embarking on interplanetary exploration. Space science and technologies offer ways to deepen human potentialities. They stimulate the best and the brightest among young people because of the challenges of developing new technologies. These include not only the fields of physics and astronomy but also include medicine and tele-medicine, genetics, biology and immunology, pharmaceutical science, psychology, and robotics. The new frontier could help each of us, confronted with the vastness of space, to identify with a global community beyond family, clan, and nation. It could aid the overcoming of nationalism, sectarianism, fundamentalism, and racism. At the same time, when different cultures cooperate and trade, they tend to expand the opportunities available to all individuals. A blossoming world literature, the printing press, the internet, and cinema around the globe are exemplary cases in which cooperation and trade have made the various different countries more creative, thus increasing diversity.

6.3 Conclusions

This book has told a story that has soared from the distant past to an equally distant future. It has depicted space exploration as a natural part and extension of the human experience. It began with horse domestication, through the wheel's invention, and the use of wind to travel faster and to reach previously inaccessible places on Earth in a reasonable time. Science is an integral part of this experience, since technology results from scientific developments. Existing knowledge and technology already enable humans today to explore the solar system. As we have tried to make the case, for humans to go beyond the solar system requires yet more advances in science and technology. We believe that the discovery of one or more habitable planets within 5–15 light years from Earth will be the crucial driver for interstellar exploration and colonisation. If the existence of a habitable planet capable of sustaining life is confirmed by astronomical observations, interstellar exploration and colonisation will raise strong interest at all levels of society, and could induce a change of heart so that governments and the private sector fund programmes for the development of interstellar propulsion technologies. Robotic exploration would still continue to play a crucial role in the early phase of the exploration of extra-solar planets in order to achieve orbital-surveillance capabilities of the targeted planets. This should provide the necessary information for the second phase of manned exploration and colonisation.

But our great remaining challenge is to overcome the paradigm of the competition for power among nation-states and, instead, to open a new frontier in space through international cooperation for the benefit of the entire human community. Cooperation in space during the exploration of the solar system would certainly produce a more stable world. Exploration and colonisation of extra-solar planets require a civilisation in which human beings see themselves as inhabitants of a single planet and global governance is conducted on a cooperative international basis. What will it take to achieve a stable world order among independent nation-states? Following H. Bull,[7] at least four conditions are necessary to achieve such stable world order. The first is a consensus on common interests and values, which would provide the foundation for the second condition, a set of rules accepted by all nations for peaceful coexistence. The third is justice. The fourth condition is a cosmopolitan culture.

The key question is whether our present civilisation is ready to embrace a culture of non-violence. Definitive answers do not yet exist. Human folly and other vices and fears can change the course and results of the journey we have tried to imagine, and so postpone such human space exploration to an indeterminate future. The analysis presented, without pretence of completeness, indicates that the basic elements for a stable world order among independent nations do not yet exist. But we want to close this book on a note of optimism. Culture is shaped by modern science. Although people have the mistaken idea that science is separate from culture, science is an important and vital part of culture, and space exploration is one significant consequence of scientific advancements. This book has sought to make a convincing argument that any space mission involving the resettlement of people will have significant economic, commercial, and cultural side-effects. Space exploration and the colonisation of the solar system should thus be on the political agenda of spacefaring nations as an important tool to strengthen world political stability, economic growth and welfare, and human potentialities and the

---

[7] Bull H. (1977).

evolution of its consciousness. Economics, sociology, political science, history, and cultural studies are thus equally important to understand space exploration and how it can contribute to humanity's progress.

Key elements of these arguments are: (i) the role of science in transforming our societies; (ii) the meaning of freedom and democracy in a world dominated by a few powerful political entities competing for finite resources; (iii) the role of the private realms of society in shaping a less confrontational world; (iv) how and why human beings learn to cooperate; and (v) the contractual basis of rights leading to tolerance in public and private spheres and to the organisation of nations in self-governing societies with democratic institutions. A last argument emerges from the activities and experiences of several organisations, namely the possibility of politics without sovereignty. Political organisations, such as Amnesty International, Doctors without Borders, and Greenpeace, and more general social movements, are active world-wide. They include groups promoting international space cooperation. The International Academy of Astronautics and the Space University are gathering scholars, space-agencies members, and informed individuals to promote and advance international space cooperation. These groups self-consciously adopt anti-statist, global identities. However, unequivocal answers to "politics without sovereignty" do not yet exist. Many think that a prosperous, valuable human life is only possible when people form communities based on allegiances to territorial states. But many others challenge this conclusion.

Answering the above questions is not within the scope of this book. But it is the authors' hope that this book will stimulate scientists and scholars of the humanities to come together in challenging conventional social, political, and economic wisdom. The central problem is how to extend ethical concerns to ever widening groups of people. In the past, we have learnt to extend these concerns to our families, our tribes, and our nations. How to extend ethical concerns to all humanity is the next challenge. At the same time, a realistic biology of the mind, advances in physics, information technology, genetics, neurobiology, engineering, and chemistry of materials are all challenging the basic assumptions of who and what we are, and of what it means to be human, thereby contributing to our evolving qualitatively in both mind and spirit.

Science and the humanities coming together, and the enlargement of the discussion to the informed general public, may help us to discover ways to form a global society that is less confrontational and more based on mutual respect and non-violence. This is not a utopian dream but a search for value systems rather than shared beliefs that could result in a more varied and hopeful world, and lead humanity to reach the stars in a not so distant future.

Reference

Bull H. The Anarchical Society: a study of order in world politics. London: Macmillan Press Ltd.; 1977.

Index

Note: Page number followed by 'n' refers to footnotes.

G. Bignami, A. Sommariva, *The Future of Human Space Exploration*,
DOI 10.1057/978-1-137-52658-8

M

R

S

Printed in the United States
By Bookmasters